Sensing Sound
Evolutionary Neurobiology of a Novel Sense of Hearing

Bernd Fritzsch

Department of Biology
University of Iowa, Iowa City, USA

CRC Press
Taylor & Francis Group
Boca Raton London New York

CRC Press is an imprint of the
Taylor & Francis Group, an **informa** business

First edition published 2024
by CRC Press
2385 NW Executive Center Drive, Suite 320, Boca Raton FL 33431

and by CRC Press
4 Park Square, Milton Park, Abingdon, Oxon, OX14 4RN

© 2024 Taylor & Francis Group, LLC

CRC Press is an imprint of Taylor & Francis Group, LLC

Reasonable efforts have been made to publish reliable data and information, but the author and publisher cannot assume responsibility for the validity of all materials or the consequences of their use. The authors and publishers have attempted to trace the copyright holders of all material reproduced in this publication and apologize to copyright holders if permission to publish in this form has not been obtained. If any copyright material has not been acknowledged please write and let us know so we may rectify in any future reprint.

Except as permitted under U.S. Copyright Law, no part of this book may be reprinted, reproduced, transmitted, or utilized in any form by any electronic, mechanical, or other means, now known or hereafter invented, including photocopying, microfilming, and recording, or in any information storage or retrieval system, without written permission from the publishers.

For permission to photocopy or use material electronically from this work, access www.copyright.com or contact the Copyright Clearance Center, Inc. (CCC), 222 Rosewood Drive, Danvers, MA 01923, 978-750-8400. For works that are not available on CCC please contact mpkbookspermissions@tandf.co.uk

Trademark notice: Product or corporate names may be trademarks or registered trademarks and are used only for identification and explanation without intent to infringe.

Library of Congress Cataloging-in-Publication Data (applied for)

ISBN: 978-1-138-49717-7 (hbk)
ISBN: 978-1-032-57653-4 (pbk)
ISBN: 978-1-351-01950-7 (ebk)

DOI: 10.1201/9781351019507

Typeset in Palatino
by Innovative Processors

Preface

When I woke up, my neck and nose hurt, my glasses were broken, and blood was all over my face. I was 19 and had just survived a car accident that resulted in a whiplash. After my neck vertebrae were realigned and my neck immobilized in a cast, I had time to contemplate with my neighbor our possible fate. For three long weeks we talked and shared a room, I was never able to see his face as he was lying head down while my head was in a sling. He had survived a motorboat accident and the propeller had shredded his lower spine. Our discussions  revolved around what would have happened to us if we would have become paralyzed from the neck down or, in his case, the propeller would have hit his thorax. After this experience, my will to live a meaningful life was at an all-time high. Years of gymnastics had strengthened my neck muscles enough to bring me through the misalignment of my neck vertebrae. The question was what problem could I find that would be big enough to engage me for a lifetime and small enough to make significant progress that would be meaningful for others to build on? Embedded in this self-declared charge was to find what type of research could benefit the most from my abilities, pretty much unknown to me. Finding "my" problem, defining an approach to solve it, and then working on it became my life's mission. About 50 years after I started this journey, it's time to share my insights with others, interrupted by a stroke at age 71.

When I had to decide what to study, I chose biology over medical neuroscience because I was fascinated with the brain and sensory processing and the emerging importance of molecular biology. This decision was also a conscious rejection of other possible directions: while I was fascinated by the big questions of the universe (where we came from, when did we start as humans, when will we end a given life), I realized that understanding, even a small aspect of my brain, would be a challenge big enough to captivate me and perhaps small enough to make an impact. It soon became clear that developmental biology, or figuring out how complex structures develop or evolve out of simple precursors, was the most fascinating of all things I encountered in biology, nowadays referred to as "Evo-Devo." Putting Einstein's spacetime into biology became my favorite focus that soon zoomed to become a single problem: the evolution and development of the ear, in particular, the hearing organ and associated novelties of the brain needed to make sense (sic!) of hearing in terms of sound decoding for communication. This insight was not reached by simply reflecting on it but was the consequence of almost 40 years of studying biology and completing my thesis work. 40 years after I first defined my life's work direction, it is time to put the various papers published on this subject into a cohesive perspective to be shared with others. My hope is that my insights can help the next generation to go where I could not reach, using techniques that will emerge to solve issues of which I was not even aware. Perhaps what I present could serve for some as a framework to build on and modify to move this fascinating problem, which is so central to human evolution, forward toward a deeper understanding. It is needed to solve some of the age-related issues associated with our declining sensory abilities and to engage in stimulating discussions.

My stroke disrupted my work, and my way of working was limited, luckily enough, temporarily. After months of rehabilitation, my ability and trust in my skill were built up again to continue my work. First among the persons I am indebted to is Dominique Crapon de Caprona. Dominique found me lying unconscious in my room, she continued to work with my rehabilitation and helped bring my language skills back to near normal. Dominique is not only my friend of 40 years, but she also had to drive me everywhere until I could to function on my own, overcoming a transient partial deficit of my right eye, right arm, and right leg. Among several people I had earlier informed about my condition, four people I would like to thank for their help: Karen L Elliott-Thompson, Jennifer Kersigo, Gabriela Pavlinkova, and Ebenezer N. Yamoah. All four people argued that I should keep on going with my work and I was encouraged by them to continue with my work. Finally, Kirsten Jorgensen was my speech therapist, she brought me back from being unable to speak to being able to communicate with words again. Without these six people, I would not have ever started to speak and think again clearly.

I thank many people who interacted with me and published papers together: Werner Himstedt and Gerd Manteuffel (Technical University of Darmstadt) who helped me start on the amphibian's development, to benefit me shaping my perception of sound, in particular Horst

Bleckmann, Ronald Sonntag, Ulrich Wahnschaffe, Udo Bartsch, Udo Will, Claudia Wilm, Martina Mans, Burkhard Hellmann, Ulrike Niemann and Heinrich Muenz (University of Bielefeld), Marvalee Wake (University of Berkely) and R. Glenn Northcutt (University of San Diego). It is my pleasure to thank colleagues from Creighton University (David Nichols, Tim Neary, Laura Bruce) and the University of Iowa (Dan Eberl, Albert Erives, Douglas Houston) for collaboration. Finally, my thanks to my past PhD and postdoc students at Creighton University (Adel Maklad, Sarah Pauley, Veronica Matei, Eduardo Rosa-Molinar) and the University of Iowa (Israt Jahan, Ning Pan, Tian Yang, Ben Kopecky, Jeremy Duncan). All were happy to know that I am much better after my Broca's area was nearly destroyed after my stroke.

Several people who interacted with me over the last 50 years are meanwhile diseased: Tim Neary, Laura Bruce, Walt Wilczynski, and Hans Straka, among several. Thank you all for your input into my thinking and for helping in various ways.

Thank you to all who have given me support to continue my work, it is appreciated.

Bernd Fritzsch

Contents

Preface   iii

Introduction   xi

1. **Defining Novelty in the Neurosensory System**   1

    1.1. Making new genes and assembling their products into novel pathways for innovative functions   4
    1.2. Sensing membrane stretching and evolving mechanotransduction   9
    1.3. New cells are generated by new molecular pathways   15
    1.4. Understanding metazoan evolution   23
    1.5. Defining hierarchy of events to understand cascading effects of changes of neurosensory evolution   29
    1.6. Chance and necessity in the formation of sensory neurons   42
    1.7. Lessons learned: Transforming existing molecular systems of single celled organisms to open new windows to sense the world in multicellular organisms using discrete sensory channels   49

2. **Evolving Mechanosensation for Gravity and Angular Orientation, Lateral Line, and Electroreception**   52

    2.1. Vestibular sensory neuron development requires a unique projection to distinguish brainstem from trigeminal projections and is the first hair cells to develop   58
        2.1.1. Vestibular neurons   59
        2.1.2. The vestibular nucleus: Origin and development   63
        2.1.3. Vestibular hair cells   67

2.2. Lateral line evolution seemingly coincides with vestibular
     development   71
     2.2.1. Lateral line sensory neurons: Define the molecular
            base of neurosensory formation from the lateral
            line placode   73
     2.2.2. Projections of the central lateral line projections   75
     2.2.3. Innervating lateral line nuclei in the brainstem   76
     2.2.4. Define hair cells and their polarities to define
            molecular basis of LL HCs   78
2.3. Evolution of electroreception: An early formation combined
     with multiple losses   81
     2.3.1. Electroreceptor sensory neurons depend on common
            placodal origin   82
     2.3.2. Electroreception of central projections innervate
            dorsal nuclei   84
     2.3.3. Hair cell formation depends on *Atoh1* and other
            genes to develop electroreception   87

3. **Connecting a Novel Sensory Input from the Ear to the Brain:
   How to Make New Neuronal Networks and New Connections
   of an Auditory System**   **93**

   3.1. The lungs evolved among sarcopterygians   95
   3.2. Formation of a spiracle *versus* a tympanic membrane   98
   3.3. The hyomandibular bone will be transformed
        into stapes   101
   3.4. Intracranial joint, basicranial muscle, notochord, ductus
        communicants, and basilar membrane are interacting   103
   3.5. The role of air and sound in water   105
   3.6. Evolving the ear shows loss and gain in sarcopterygians   107

4. **Defining the Auditory System of Tetrapods**   **113**

   4.1. Auditory spiral ganglion neurons are unique
        among mammals   116
   4.2. Auditory nuclei are expanded to add additional
        connections   121
   4.3. A unique set of genes are needed for hair cells
        in mammals   124

5. **Sound Processing: Boundaries for Auditory Signal
   Processing Revealed**   **129**

   5.1. Auditory neurons innervate selectively the topologically
        auditory nuclei   130

5.2.	*Atoh1* defines the cochlea nuclei as well as electroreception and lateral line nuclei	133
5.3.	Cochlea nuclei are unique in mammals	138
	5.3.1. Bushy cells (AVCN)	139
	5.3.2. T-stellate cells (PVCN)	139
	5.3.3. D- and L-stellate cells (PVCN)	140
	5.3.4. Octopus cells (PVCN)	140
	5.3.5. Root cells (PVCN)	141
	5.3.6. Fusiform (pyramidal) cells (DCN)	141
	5.3.7. Giant cells (DCN)	141
	5.3.8. Granule cells and unipolar brush cells (VCN, DCN)	141
	5.3.9. Golgi cells and superficial stellate cells (DCN)	142
	5.3.10. Tuberculoventral and cartwheel cells (DCN)	142
	5.3.11. Multimodal output and inputs	143
5.4.	Outflow of cochlear nuclei to SOC	146
	5.4.1. Lateral superior olive (LSO)	146
	5.4.2. Medial nucleus of the trapezoid body (MNTB)	147
	5.4.3. Medial superior olive (MSO)	147
	5.4.4. Superior paraolivary nucleus (SPON)	148
	5.4.5. An integrated perspective of the SOC	149
5.5.	Lateral lemnisci are generated from two distinct populations	150
	5.5.1. Ventral nucleus of the lateral lemniscus (VNLL)	151
	5.5.2. Intermediate nucleus of the lateral lemniscus (INLL)	151
	5.5.3. Dorsal nucleus of the lateral lemniscus (DNLL)	151
5.6.	The inferior colliculus (IC)	153
	5.6.1. The central nucleus of the inferior colliculus (ICc)	154
	5.6.2. The lateral nucleus of the inferior colliculus (ICl)	155
	5.6.3. The rostral nucleus of the inferior colliculus (ICr) and dorsal nucleus of the IC (ICd)	155
	5.6.4. Commissural interconnection and neurochemistry	156
5.7.	The medial geniculate body (MGB)	159
	5.7.1. The ventral division of the MGB (MGBv)	160
	5.7.2. The dorsal division of the MGB (MGBd)	160
	5.7.3. The medial division of the MGB (MGBm)	161
	5.7.4. The reticular thalamic nucleus (RTN) and the posterior paralaminar thalamic nuclei (PPTN)	161
	5.7.5. Integrated functional perception	161
5.8.	The auditory cortex (AC)	162
	5.8.1. Development of the AC	163
	5.8.2. AC laminar distribution and basic connections	164
	5.8.3. Corticocortical and corticothalamic connections	165

	5.8.4. Cortical plasticity: From the ear to the AC	167
	5.8.5. Auditory cortical processing revealed	169

6. The Evolution of the Auditory System as a Blueprint for Sensory Expansions Using Established Principles – Evolving New or Adding Existing Senses to Expand the Perception of the World   171

    6.1. Evolving the auditory central projection revealed   171
    6.2. Auditory sound processing   176
    6.3. The auditory cortex reaches out beyond the A1   179
    6.4. The role of the hippocampus to maintain hearing   182

7. Summary and Conclusion   185

References   188

Index   208

Introduction

In any overhaul of construction, a critical decision to make is how much of the old to preserve, how much to add to repurpose existing features and how much to tear down to replace with something new. This basic principle of construction overhaul can be applied to understand the evolution of a novel sensory system as an addition to an existing central nervous system. This basic insight has one major caveat: All human sensory modalities evolved in long extinct ancestors as all craniate vertebrates have all major senses of the head (eyes, ears, nose, taste), whereas all vertebrate outgroups (amphioxus, tunicates) have neither of those senses. Unfortunately, this situation leaves us with no trace to follow the implementation of novel systems beyond modifications of existing systems, best exemplified in the evolution of the retina and eye muscle system (Fritzsch and Martin, 2022).

Among the craniate head senses, hearing is the sole exception to this rule: organs dedicated to sound extraction and neuronal centers dedicated to sound information processing evolved multiple times in many animals, always through the re-shaping of existing mechanosensitive organs while developing novel information processing centers in the brain. Hearing is thus the only major vertebrate and invertebrate sense whose evolution can be traced from the molecular basis of cellular transformation to, in the case of humans, the societal impact of the fully developed sense of hearing. Moreover, hypotheses about the stepwise implementation of this sense can be tested through genetic and other experimental manipulation. Such experiments move auditory evolution from a descriptive level that is connected to evolution by inductive reasoning and inferences to the experimental level needed for scientific support or otherwise, at best, tentative and disputable ideas. Furthermore, the physics of sound is well understood, and technical applications of sound are abundant in our society in terms of load speakers, microphones, and ultrasound imaging, to name but a few applications. The purpose of this project is to provide a readable account of sound processing related changes at multiple levels, from molecules through cells to organs of the vertebrate ear and brain, and summarize the implications of these changes for human cultural evolution. Emphasis will be given to data supported by experimental evidence and critical experiments needed to support conclusions will be pointed out.

After the evolution of the auditory system is laid out, the lessons learned are explored in terms of forward looking possible sensory expansions to allow the human mind access to currently hidden sensory dimensions of the world, in part explored by senses of other animals that are alien to us such as ultra- and infrasound, magnetoreception, electroreception, ultraviolet and infrared vision. The question to be addressed is: can one use insights into the evolution of a new sense as a blueprint to implement novel senses into the human brain thereby expanding our perception of the world around us, currently limited by the restricted range of stimuli perceivable by our senses? The quest here is whether we can expand our sensory windows to the world from the current slits to a panoramic view to grasp better the world around us and ultimately enhance our understanding of the cosmos currently restricted through evolutionary adaptation of senses and the associated brain structures to ensure survival on this planet with his current set of physical and biological constraints?

All this analysis is placed in the context of the scientific progress we have seen since Giordano Bruno was burned at the stake for arguing against the dominant world view of his time, still held as a dogmatic view by far too many, despite the enlightenment of the last few centuries. Since Bruno, Galileo, Copernicus, Newton, and others have changed the scripted world view, most educated people know that the solar system we inhabit is but a minor star at the edge of a small galaxy floating somewhere in a gigantic universe that is ~14 billion years old. We know that simply because it takes time to burn all hydrogen into helium and fuse helium inside early suns into the larger chemical elements needed to make planets that can sustain organic life as we know it on Carbon, Oxygen, Nitrogen, and others discovered only some 150 years ago and nicely summarized in the famous Periodic table of Dmitry Mendeleev. Ever since Ludwig E. Boltzmann's laws of thermodynamics were put into the emerging context of the evolution of our universe it has become clear that complex molecular assemblies needed for life are Nature's way of increasing the entropy in the universe. Importantly, Boltzmann (Boltzmann, 1887) was the first to realize that the physical properties of matter (liquid, gas, solid) arise from the properties of atoms (chiefly among them are charge and mass). What is most fascinating being the transitions of matter (from gas to liquid, from solid to liquid, from solid to gas) and the conditions defining those transitions. As much as the transition of liquid water into ice can happen in a fraction of a second while each prior state lasts forever given the defining parameters (pressure and temperature; Fig. 1). Boltzmann's insights into the logarithmic connection of entropy and probability, stated in the kinetic theory of gases are summarized in his $S=k_B \ln W$ (S, entropy; k_B, Boltzmann's constant; ln, natural logarithm; W, probability [Wahrscheinlichkeit]). While the phase transitions follow as a macroscopic cohort effect directly from this formula, it also implies that individual particles (atoms or molecules) have microstates that can be anywhere in that phase transition, best exemplified in the triple point of gas/liquid/solid or in the state of supercritical fluid. This probabilistic outcome at a single particle is difficult to understand and I like to provide another attempt here to explain this.

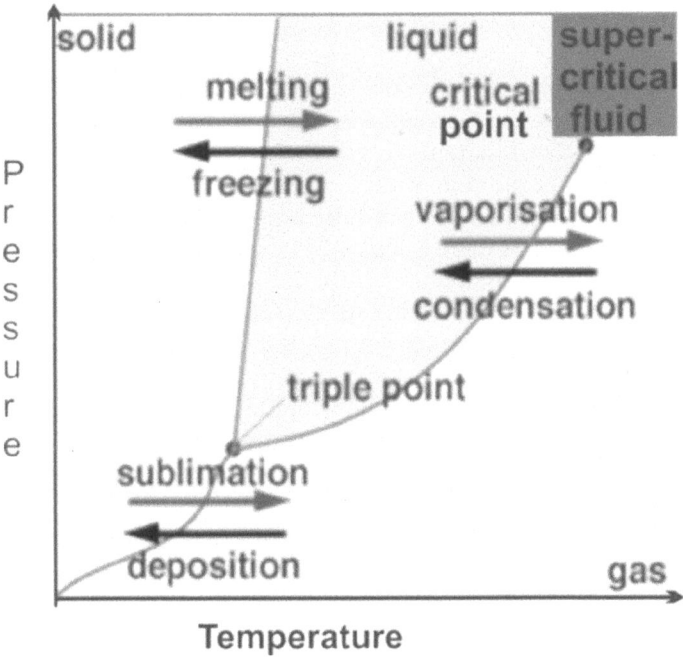

Figure 1: General phase change diagram shows the four states of matter and the transitions that depend on temperature and pressure for an ideal gas.

Suppose you have all the letters on this page in your hands and can throw them on a piece of paper. How likely is it that they end up forming readable words, whole sentences, or fall exactly in the lettering, including spaces, as displayed here? Not even close that likely will happen, correct. However, suppose you do this for the next 4 billion years every second. By random chance, you will end up with rare occasions resembling closely what is written here. If you take any approximation and put in the work to arrange those letters you can copy what is written. This provides the basis for considering how life evolved, a phase transition from inorganic to organic that is now maintained by extracting energy from other sources, increasing their entropy while keeping the somewhat orderly assembly of molecules that make up living beings alive. Obviously, for such systems to 'live' they need to generate energy out of processes that increases the entropy around them. Simply speaking, feeding humans is from an entropy perspective identical to burning the crops on the field: either way we destroy the solar energy stored in complex biomolecules into heat, carbon dioxide, and water. Humans and other heterotrophic organisms can simply be regarded as entropy maximizers while green life uses solar energy in the form of photons to decrease their entropy, building complex macromolecules that store that energy. Ultimately, all life depends on the solar fusion reactor that is driven by the Einstein insight ($E=mc^2$) to provide most of our 'life-enabling' energy, the sun.

As with all phase transitions outlined above (Fig. 1), the transition from inorganic molecules to organic life will require precise definitions of the conditions, and therein lies the problem: life on our planet may be over 3 billion years old and what conditions prevailed at that time is a matter of speculation. One theoretical aspect of Boltzmann's insights is that given enough time there is a certain random chance that a 'Boltzmann brain' spontaneously aggregates through a series of improbable events (much like all letters fall to formulate all written on this page!). This book is not about the possible formation of a Boltzmann's brain, but attempts, using known biological principles shaped by the physics of our planet, to explain the neurosensory evolution of a mammalian-type hearing system.

CHAPTER 1

Defining Novelty in the Neurosensory System

Nothing Comes from Nothing *(Parmenides, ~480 BC)*

Biology is the master of innovation within a framework of historic constraints, tinkering with existing genetic information to developing novel forms, evolving, among many astonishing beings, the human species with its pioneering mind, able to generate vast societies benefitting from revolutionary technologies. As an example, the novel communication platform offered by modern cell phones has, in only 10 years, transformed societal communications radically. Such effects were not foreseeable when the first commercial Apple computer was rolled out in the early 1980s, a mere 20 years before the smart phone. Going beyond these technical innovations, understanding novelty in Biology requires defining the boundaries set forth by previous innovations, now embedded in the inherited genetic blueprint, constraining the freedom of innovative space usable for Biology. These inherited limitations, handed down by the biological information storage system of the DNA, seem to add, at first glance, an additional burden to novelty in Biology compared to the technical world. In the human generated technical world, one useful design may be replaced by radically different designs, maintaining only the functional principle but none of the structural details of previous models. In contrast, biology tinkers with the development of existing structures to achieve novel functions, leaving remnants of past solutions hidden in the apparent novelty, traceable by modern genetic analysis.

For example, the propeller-piston engine of earlier airplanes was replaced by a different kind of engine, the jet engine, lacking pistons and propellers altogether. Even more different are the rocket engines that can bring man to the moon and back, retaining only the concept of thrust underlying all engines. In contrast to these very different engines that provide thrust to keep human flying machines in the air, Biology would find ways to morph a propeller into a rocket engine. This ability to tinker with existing structures is particularly obvious regarding the wings that propel birds, bats, flying fish, and extinct pterosaurs. All flying vertebrates transform forelegs/fins into different shaped wings that all use the same basic principle to generate both lift and thrust, a principle that was only very recently mastered by humans beyond soaring with stiff winged airplanes. At a molecular level, bird wing and mouse leg development start very similarly, and this similarity is shared with all four legs, whether they develop into fins, limbs, or wings.

This example highlights that engineers are in principle, only limited by their ingenious implementations of techniques, constrained only by the law of physics. In contrast, biology is limited by its history, inherited via ancestral genes that require transforming forelimb developmental programs to achieve a different outcome for flying such as bats, birds, flying fish, and pterosaur wings. Any deep understanding of the boundaries of biological innovation must therefore decipher the inherited transformation of the genetic blueprint that governs cells, organs, and organismal development. Combined, this can elucidate how novel morphologies such as a bat or bird wings are shaped by both physics and molecular constraints to allow flapping of wings for active flying, including lifting off and landing under various conditions, through modification of ancestral forelegs.

This book will explore the evolution of molecules needed to transduce mechanical stimuli in single cells prior to the evolution of metazoans that eventually specialized for sound detection in vertebrate hair cells. Hearing was chosen as it is the ONLY major vertebrate sense whose evolution can be traced among living vertebrates as a transformation of sensory epithelia in the ear through accessory structures into dedicated sound detecting organs. These sensory epithelia in the ear, dedicated to hearing, evolved novel neuronal connections to unique centers in the brain specialized for the processing of sound. The combination of dedicated ear sensory epithelia combined with faithful sound processing centers, able to reliably decode messages conveyed by sound modulations, provides a unique example of evolutionary developmental neuroplasticity that remains largely unexplored in contemporary theories of neurosensory and brain evolution. This book will provide a first attempt to correlate the innovative evolution of sound processing in the context of evolving a brain with enough neuroplasticity to enable human's ability to comprehend modulated sound that is the basis of human language evolution.

Without a doubt, hearing is a prerequisite for the evolution of language and thus the development of human societies able to relate insights gained by one generation of an ancestral human population in oral and, today, in written form to the next generation, developing a non-genetic 'inheritance' of insights via narratives and scriptures. Beyond the historic importance of auditory plasticity in the context of evolving human language and thus our societies, this book recognizes the capability to use that language to generate a novel domain of narrative, not restricted by the short term processing of information, to survive. I posit that this alteration of the time domain of information storage provided humans with the ability to generate innovative ideas that enabled their cultural evolution beyond biological limits that bind other species. This human ability to use extra-genetic information storage empowered humanity to evolve, over only 10 generations (the last 200 years), an engineering and art driven society that is now defining the future of men and our planet. The risk and benefit of this uncoupling of what I refer to as 'literature space' from 'biological space' (the 'reality' provided by all other sensory inputs) are providing at the same time the greatest opportunities and the utmost risk for life on our planet (Club of Rome: https://www.earth4all.life/). Whether the brain evolution that enables us to destroy our world is growing faster or slower compared to our understanding of and controlling the unintended impact of us on our society, remains to be seen.

The thesis proposed here is that all these achievements derive from the necessity to evolve neuronal plasticity that allowed learning how to extract meaning from human generated sound as languages are not encoded in our genes, only the ability to acquire language. This could only be accomplished by evolving a plastic system able to learn novel meanings to any given sound modulation possible to men through plastically reorganizing existing connections and establishing new ones. It is noteworthy that man-made machines have now implemented similar learning strategies using brain-wiring inspired chips. Equipped with self-teaching algorithms such as computer based Artificial Intelligence (like OpenAI) can meanwhile outcompete humans in sophisticated games such a *Chess* and *Go* and can understand language at an increasingly sophisticated level, including 'reading' brain waves underlying words. This book illustrates how life could have evolved the brain's ability to generate learning mediated changes that allowed humankind to construct such AI. My thesis is that in generating AI we have essentially copied our own success story, necessitated by the evolution of learning abilities to decode the meaning of sound, and evolve human communication, like ChatGPT.

Deconstructing the evolution of the human brain over the last 200 years was driven by a naturalist perspective to analyze certain steps deemed important and accessible to analysis based on the current or past technical limitations such as volume increase of the brain. The underlying hope was that by analyzing the parts the meaning of it all might emerge. While such emerging properties on a smaller scale have certainly driven much of our scientific insights, I posit that large scale emerging properties require a system level analysis in addition to understanding the necessary details. Perhaps the differences in studying small- and large-scale emerging properties can best be exemplified using progress in computing over the last 60 years which also highlights the impossibility to deduce large scale emerging properties by simply quantifying size without an understanding of computing power of a given unit.

Over the last 60 years computing power and size differences in computers have gone in opposite directions: a 1960 PDP 11 mainframe was filling an entire room but had far less computing power than a modern smartphone. Likewise, we now know that the packing density of neurons varies dramatically and allows parrots with less brain volume to pack as many neurons as a cat into a much smaller volume (Herculano-Houzel, 2016). Instead of engaging in functional speculations of brain size variations, I attempt here a reconstruction of evolutionary modifications leading to the formation of the necessary features to accomplish auditory signal acquisition and analysis, including learning induced plasticity. Such an approach allows tying all steps toward accomplishing this into the physical properties of sound and sound related mechanical stimuli. In essence, it will allow following the logic of a molecular implementation of changes, reconstructing a plausible trajectory towards the generation of novel structures followed by an analysis of possible selective advantages. Understanding the properties of the system or not is a large-scale emerging property of detailed analysis at the microscopic, cellular, and molecular level (bottom up) but rather follows from an understanding of the physical properties of sound to drive the analysis of the details of the mammalian sound analyzing system (top down). With this approach in mind, I will start first to understand the evolution of the necessary parts.

1.1. Making new genes and assembling their products into novel pathways for innovative functions

Following the initial insights of Ernst Haeckel (Haeckel, 1876), that cell nuclei are the physical entities of inheritance, and the discovery of Friedrich Miescher of 'nuclein' (now deoxyribonuclei acid, DNA) in the 1860ies (Glover et al., 2018; Miescher, 1869), it took almost 100 years before the significance of these findings were clarified. James Watson and Francis Crick, using the crystallographic insights into the molecular details of DNA generated by Rosalind Franklin (Franklin and Gosling, 1953), developed a model of DNA structure in the early 1950s, proposing the possibility of DNA being a biological information storage device. Succeeding this first insight, we have now completely sequenced the human genome in 2000 and begun to understand not only protein coding genes but also the regulatory capacity of the non-protein coding part of the DNA. We begin to understand the 2% of the human DNA that is coding for proteins but our understanding of the 65% of DNA that is forming long non-coding RNA (lncRNA), or microRNA (miR) is still rudimentary as is our understanding of regulatory elements that drive differential gene expression (Shafin et al., 2020). While the importance of controlling elements that allow the regulation of transcription has become clear, the detailed understanding of this regulation is nowhere good enough to allow entering this into an evolutionary scenario (Peter and Davidson, 2015). Importantly, because of novel techniques developed to sequence the human genome, it is now affordable to sequence various human genomes for medical purposes and figure out how even single nucleotide changes or short duplications and deletions affect genomic function. Third generation sequencing makes it affordable to sequence numerous other species thus generating a complete data base of the genomic evolution of extant life using new algorithms that allow arranging genes according to their sequence similarities into groups of related genes and annotating the functional space covered by such molecularly related proteins. In special cases, we can now trace the same highly conserved gene from single celled ancestors to humans and can define the molecular changes in terms of their functional significance (Zoonomia. Vignieri, 2023). Modern technology allows storing entire human libraries in modified DNA, in analogy to the libraries imprinted by evolution in the DNA, thus validating the idea of Watson and Crick about the true function of DNA as a trans-generational information storage device.

All these insights over the last 150 years added molecular detail to the logic of inheritance described initially by Gregor Mendel. We now know that inheritance follows a semi-conservative pattern: one of the two double strands of DNA is retained whereas a complementary new strand is synthesized to provide the daughter cells with the parental information. This process, combined with the sorting of different DNA molecules (the chromosomes of eukaryotic cells) during cell divisions and the mixing of parental DNA during crossing over as part of Meiosis to generate sperms and eggs, result in minor or major errors that affect the next and all following generations. In essence, DNA is a trans-generational 'hard drive' that brings parental information on how to build and maintain a functional organism to the next generation, nearly faithful, but with minor 'errors' (we call 'mutations') that can lead either to dead ends or to innovative new structures and modified functions. This ability of DNA to undergo minor changes from generation to generation provides its **'evolvability'** (Wagner, 2011). Moreover, this process ensures that children inherit a mix of maternal and paternal genes, providing a novel combination of various genetic differences not found in either parent. Children are not clones of their parents but unique novel entities. Mutations and the mixing of parental genetic information ensure **the arrival of the fittest** (Wagner, 2015) on which selection can act to ensure the survival of the fittest.

In humans, this 'error' rate in genomic inheritance was quantified by comparing the entire genome of parents with that of their children. The data suggest that around 100 copy mistakes are made per generation. A child is thus not only the random mix of both parental information, with 50% genetic information coming from the mother and father (less from the father for a boy because of the low genetic information of ~100 genes on the Y chromosome, compared to ~1300 on the X chromosome in females) but also is unique through these novel mutations at a tune of 0.01-0.001% of its genetic information (Tan et al., 2014). This copy-error rate (or mutation rate) is within that order also in other organisms, meaning that all living organisms always generate offspring that is a mix of both parents with an added extra of novel mutations. Moreover, detailed analysis has demonstrated that the DNA of each species is composed of 'hot' and 'cold' spots with a higher or lower probability of mutations. Simply speaking, mutations are not fully random but show aggregations that are not fully understood in their genomic background.

Independent of the generation cycle (in humans ~20 years, in yeast several hours) these mistakes cause alterations of the DNA that result in differences between generations. Obviously, the faster the generation cycle the faster these alterations (or mutations) will build up over identical time spans, consequently changing protein composition of cells and thus cells. Combined, these mutations provide the raw material on which selection acts to filter out the most suitable combination of proteins making an organism survive and propagate in each environment, following a Darwinian selection to evolve towards a better adaptation to a given environment with time. Simply speaking, *genetic variations provide the arrival of the fittest whereas selection picks the survival of the fittest* (Wagner, 2015). These basic principles have long been undisputed truisms of biological teaching and form the material basis for the **Theory of Evolution**, initially proposed by Charles Darwin around the same time that all the other relevant discoveries around the inheritance of characters occurred, but over 100 years before the mechanistic explanations of mutation were understood. It is only in the last few years that our comprehension of the mechanisms driving variation of genes and their encoded proteins has reached a level that allows defining mechanistically trajectories taken by living cells from their humble beginning some 3.5 billion years ago to generate the complex senses and brains of vertebrates, including humans, that will be explored here (Peter and Davidson, 2015; Vignieri, 2023).

Another important aspect of evolution that confuses is the number of genes and amount of DNA each organism has. Naively one would expect that somehow complexity of life correlates with the amount of DNA to store the necessary information coding for the development of a complex organism, like humans with over 200 distinct cell types, as compared to single cell organisms. While this assumption is true for the distinction of bacteria and multicellular organisms (1–10 million base pairs versus 100 million to 100 billion base pairs in eukaryotes) it does not hold up within metazoan to protozoan comparisons. Humans have about 3.2 billion base pairs but some salamanders, some plants and some single celled amebae have over 100 billion base pairs or over 30 times the amount of DNA to store their genetic information, clearly much more compared to humans. Simply speaking, neither the total amount of DNA nor the number of proteins encoding genes correlate in a simple way with what would typically be considered by men as morphological complexity. Unless one argues that ameba, a single celled organism with no stable shape, is the most complex organism there is. Or that salamanders and lilies are more complicated compared to humans?

With this paradox in mind, there is, however, a correlation between genome size and complexity in most metazoans with genome doubling at the base of metazoans and chordates being a likely explanation for the multiplication of genes that allow the selection of different mutated forms of 'sibling' genes that arose through such duplication. Indeed, duplication of existing genes followed by sub- or neo-functionalization (or mutational dysfunctionalization into pseudogenes) has long been recognized as a major driving force of evolution. As Ohno pointed out (Ohno, 1970, 2013): *"In a strict sense, nothing in evolution is created de novo. Each new gene must have arisen from an already existing gene."* **All genes came about through tinkering with the original DNA of the common ancestor of all living on earth for the last 3 billion years.**

In the absence of any understanding of the physical nature of inheritance, contemporaries of Darwin could still speculate whether acquired characters could be inherited. As naïve observers, lay people will rightfully claim that humans are immutable, as they will hardly witness massive changes in humans over their lifetime, while considerable changes due to mutations typically will be dismissed as aberrations (mutants) that are less adaptive. Either position is equivalent to proposing a flat-earth because one cannot see the curvature of the planet earth without additional simple understandings (such as comparing the length of a shadow by a stick of the same length at the same day and time of the year but separated by ~1000 km in North-South direction) and reflects the limitations imposed on our perception by our senses without an appreciation of the true physical limitations imposed by our universe. Darwin proposed his ingenious idea of evolution at a time when neither DNA and its distribution in nuclei only, nor the mode of inheritance proposed by Mendel, nor the details of transcription of DNA into RNA followed by translation into proteins were known. Darwin's bold hypothesis has seen ample evidence over the last 150 years that has clearly established evolution as the guiding theory to understand the diversity of life and its adaptation from simple beginnings, thus providing the basis for modern biology. As T Dobzhansky put it so well: **"Nothing in biology makes sense, except in the light of evolution"** (Dobzhansky, 2013).

One major insight generated by the molecular revolution over the last 50 years is that all life on earth, from bacteria to humans, use essentially the same molecular code to specify amino acid sequences in proteins out of the nucleotide sequence of the DNA. This fact is most parsimoniously explained by assuming that this complex interplay (DNA codes RNA codes Protein) evolved only once and was inherited from a single universal ancestor of all life, the **LUCA** (**l**ast **u**niversal **c**ommon **a**ncestor). While we still do not fully understand how LUCA came into existence some 3.5 billion years ago, we know that this ancestor must have had a single strand of DNA that coded for the proteins using the same nucleotide code we still use after millions of years since LUCA to translate the RNA message transcribed from the DNA into proteins. In fact, this universal code found in single celled organisms and humans alike strongly supports inheritance from a single ancestor living over 3 billion years ago. Alternative explanations require the action of yet to be identified principles that shape DNA/RNA/Protein to interact in exactly the same way using seemingly identical proteins that function alike in unicellular and multicellular organisms, a position that requires an extremely unlikely set of assumptions that make it less parsimonious compared to the single, unique life origin hypothesis. There is a very slim chance that yet to be understood constraints may have multiple times in parallel generated molecular identical life but overcoming this improbable beating of odds would imply an extremely strong selection of only such pathways, against all others. For example, a technically more elegant solution would have been to use three base pairs (not the currently two pyrimidine and purine base pairs of adenosine, thymidine, guanosine, and cytosine) to generate, already at the level of the DNA, the triplet code used by RNA to identify a given amino acid to be incorporated into the growing string of a protein. Likewise, it is unclear why life on our planet limited itself to (mostly) 20 amino acids of several hundred to build peptides. Conceivably, up to 60 amino acids could be differentially encoded using the triplet code (4^3=64) which would render it less redundant. Combining 6 (instead of 4) nucleotides with the same triplet code would result in coding for an even larger number of amino acids (6^3=216). Why the 64-possibility code evolved instead of the 216, and why only ~20 amino acids are encoded and not 60 are major issues of the early evolution of life that are unresolved (chance or design). It remains also unclear how the transition from complex macromolecules to life progressed. Multiple hypotheses have proposed of RNA first or metabolism first. Clearly, none of these competing hypotheses have enough evidence to overrule others unless life can be regenerated in the dish. The physicist Feynman famously remarked: **"What I cannot create, I do not understand"** (Samantsidis et al., 2020). We must admit that we currently cannot generate life out of macromolecules and thus in a sense do not yet understand the transition from free floating macromolecules

Defining Novelty in the Neurosensory System 9

into a membrane bound compartment of cellular life. NASA in its search to identify life on other planets within and outside the solar system has defined life as: "A self-sustaining chemical system capable of Darwinian evolution." This definition implies a system that maintains internal order across generations by absorbing complex energy (such as sugar) from the environment and emitting simple molecules (CO_2, H_2O) while generating internally useful energy (on earth it would be ATP out of ADP).

1.2. Sensing membrane stretching and evolving mechanotransduction

By necessity, the DNA/RNA/Protein mix of the first cells must have been separated from the environment through a lipid bilayer to avoid diffusion of the macromolecules. This lipid bilayer evolved into the cell membrane, generating a micro-bioreactor able to use existing molecules to fuel its own existence, including its own growth and propagation. The absence of such a lipid bilayer would have resulted in the dilution of the cell specific mix of DNA/RNA/Proteins. The specific mixture resulting in the origin of life on earth as we know it would have been diluted and lost. Many theories have proposed how the first living cell may have come into existence and how much of this was due to chance and necessity. Billions of planets have been found around suns in our own galaxy, many of them possibly billions of years older than the earth, and many with conditions suitable to sustain life as we know it in the 'goldilocks' zone that allows for liquid water on a planet's surface (not too close or too distant from its star). It stands to reason that we will eventually find out whether life was a unique chance event that only happened on earth, not repeatable on other planets due to an unclear fortuitous combination of yet to be defined conditions. Alternatively, forming life is driven by physical and chemical laws and thus is ubiquitous on every planet able to sustain it in the proper environment, much like most suns seem to be orbited by planets. Some arguments can be made that the transition of organic molecules to life was driven by Boltzmann's laws of entropy.

Imagine a blue sky. Obviously, there is only one possible way to make such a blue sky and thus the entropy of this can be considered low. Now imagine a completely overcast sky with amorphous clouds that blend into each other. Obviously, to go from the blue sky to completely overcast requires the water vapor in the air to increase and form clouds that increasingly cover the sky. The universe, and thus the earth, is in the middle of the transition from the big bang with high entropy to the final demise with low entropy. Much like the transition of the universe generates the various galaxies including our Milky Way or Andromeda you can imagine that clouds have certain features in common but their individual shape, how they blend into each other, and how fast they cover the sky will depend on initial conditions that could be modified by the beating of a butterfly wing. Life as we know it is just another direct consequence of the transition phase the universe and our planet are currently in. If so, logic would require that life as a self-sustaining system will form on every planet within the right zone and right ingredients, driven by one or two photons to absorb light (Pawlicki et al., 2009). Readers interested in this subject should look up modern references on cosmology and the evolution of life listed under further reading. I will be focusing here only on DNA/RNA/Proteins that are inserted into the lipid bilayer that covered LUCA, separating the internal part of this original cell from the surrounding world, not how this first cell came into being.

The lipid bilayer surrounding a given cell is flexible or pliable to mechanical deflection across its major plane but is stiff with limited flexibility to movement within the plane of the membrane such as exerted by cellular volume changes due to osmotic changes. Cell volume changes are caused, for example, by metabolizing larger molecules into multiple smaller ones, increasing osmotic pressure inside the cell relative to the outside that could burst a cell. Being metabolically active to maintain its own low entropy while increasing the entropy of its environment is an essential step towards life: Life depends on catabolizing large organic molecules such as sugars into smaller components, such as H_2O and CO_2, to generate energy. This process will affect the osmotic pressure in a given cell, necessitating the evolution of membrane spanning proteins that can form safety valves as one very early step in cellular evolution. In order to equilibrate molecular density and osmolality outside and inside a cell, water will leak into such a cell via aquaporin-like membrane spanning proteins but also through ionic leaking with water surrounding ions. Incoming water causes cell volume to increase with the possibility to disrupt the lipid bilayer to spill the content of the cell into the environment unless countermeasures are taken. To avoid such catastrophic effects, single cell organisms have channels inserted into the lipid bilayer that act like safety valves. These 'safety valves' open upon lateral stretching of the lipid bilayer because of volume increase or other forces acting on the lipid bilayer to allow equilibration of the cell's content with the environment (Fig. 2). Valves will shut down when the volume of the cell and thus the stretch force on the membrane and its protein content will decrease. In addition, ionic flux and pumping of Na^+, K^+, and Cl^- will also affect the cell volume, indicating that the earliest single celled organism in all phyla may have had already water and ionic pumping mechanisms later associated with sensory cells and neurons.

The best-understood channels in terms of molecular modeling are the tension gated mechanosensitive channels in bacteria. These channels

Defining Novelty in the Neurosensory System

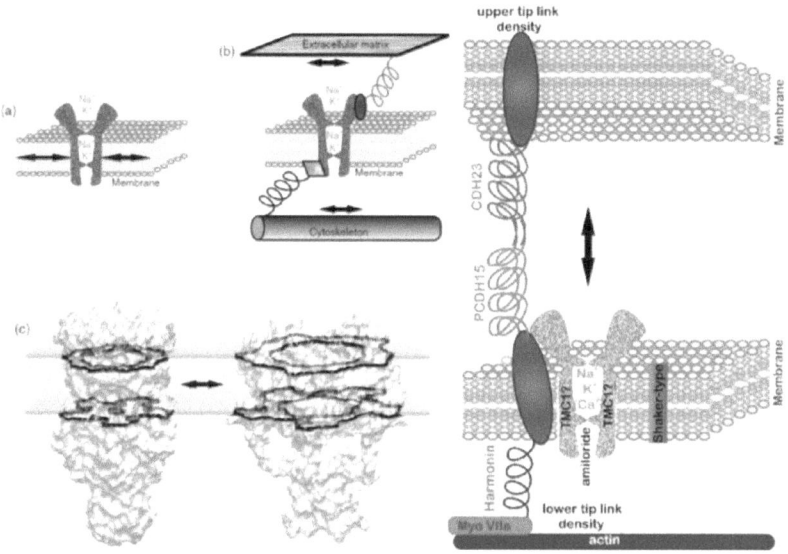

Figure 2: Various channels that respond to changes in turgor resulting in stretching of the membranes (double arrow in (a)) have been found in single-cell organisms. It is possible that such mechanosensitive channels (Msc) were modified in the unicellular ancestor of metazoans through extracellular or intracellular matrix attachments to provide increased sensitivity for shearing forces. Molecular evidence suggests that, across metazoans, family members of degenerin/epithelial sodium channels (DEG/ENaC) and transient receptor potential (TRP) channel families and transmembrane channel proteins (Tmc's) are candidates for Msc. It thus remains possible that metazoan ancestors selected either or both families for specific properties that allow increased sensitivity to detect mechanical stimulations. What such properties could remain unknown in the absence of any model of the sensitive mechanosensory channel in any metazoan taxon. Detailed models of the pentameric large mechanosensitive channels (MscL) of bacteria suggest an iris-like opening upon tension acting in the plane of the membrane (c). In vertebrates there is no extracellular matrix or cuticula connection. Instead, two stereocilia are interconnected presumably by *Cdh23* /*Pcdh15* which is hypothesized to be anchored to MyoVIIa via harmonin. Loss of any of these genes results in deafness indicating that in vertebrates mechanosensation requires relative movement against the actin core of the stereocilia. Additional connections exist between *Cdh23* and Myo1c, but no knockout data support the claimed function as an adaptor. It is speculated that MyoVIIa transports the still unknown amiloride-sensitive mechanosensory channel to the tip, but it is unclear whether this connection remains past development (dotted line). In nematodes, at least two essential subunits of the DEG/ENaC channel are known whereas it is not clear whether the vertebrate channel is composed of DEG/ENaC or TRP channels. Certain candidates have been excluded as mutants in, for example, TRPA1 do hear excluding an essential role of this protein in mechanosensory transduction. (Modified after (Fritzsch et al., 2020)).

respond to the tension of 5–10 dyn/cm (or 5–10 mN/m) along the membrane with an iris-like opening at a time scale of 10 ns. Importantly, such gating times could accommodate even frequencies of over 100 kHz, which is as rapid as transduction in mammalian cochlear hair cells, indicating that mechanical gating of such pores has emerging properties predate by millions of years those of mechanotransduction channels of vertebrates that serve hearing at very high frequencies beyond our hearing range. While the gating properties of intramembranous bacterial channels are within the range of vertebrate mechanosensory channels, their sensitivity is several orders of magnitudes less. Tethering vertebrate mechanosensory channels to either part of the intracellular cytoskeleton, or to processes protruding from a cell, or both intra- and extracellular tethering should increase the sensitivity to reach that of the vertebrate mechanosensors. Vertebrate mechanosensors have been calculated for hair cells to perceive displacements of a few nanometers with forces of <10 pN/m, several orders of magnitude more sensitive compared to the bacterial membrane pores (Fig. 2) thanks to the tethering to adjacent cellular processes. Of course, transmembraneous pores of single celled organisms could be combined with and gated by chemical, or photic or even electric stimuli, allowing a given free living cell to respond to a variety of signals with the limited repertoire of motor responses such cells can. Some of these receptors are highly localized in single cells allowing sensing direction or light or of gravity, implying that subcellular localization of various transducing systems is already accomplished in single celled organisms.

The above paragraph has clarified that the evolution of a membrane embedded channel protein that can respond to the mechanical stimuli in the frequency range of mammalian hearing could have predated the evolution of mechanotransducing cells of metazoans able to decode sound signals by millions of years. One major obstacle confronting vertebrate mechanotransduction was that the molecular nature of the tethered channel in hair cells was unknown until very recently and thus molecular ancestry of the channel protein could not be studied. Available data now strongly implies that the channel is composed of *Tmc1/2* proteins, with the possible addition of other channel proteins such as *Cib2*. Previous work that implied the nociceptive channel TrpA was disproved, but the TrpA channel may still play a role in high intensity sound perception in the ear leading to pain sensation. Despite significant progress toward revealing the molecular composition of the vertebrate mechanotransduction channel, its true configuration as well as the association with the extracellular and intracellular tethers via well-characterized proteins (*Cdh23*, *Pcdh15*) is still unclear. In addition, possibly not all elements forming the mechanotransduction complex are known (Fig. 2). Surprisingly, the details of mechanosensory channels seem to vary even between hair cells of the same organism. For example, mechanosensory function and even viability of hair cells in mammals depend critically on *Cib2* and yet the vestibular hair cells show no detectable functional decline nor altered viability after *Cib2* is deleted. At the very least, this indicates that properties of the mechanotransduction complex can be modified through the addition of another component in different hair cells in the ear of the same species adding to the complexity of elucidating the molecular history of the vertebrate mechanotransduction channel. Outgroup comparisons with non-vertebrate chordates such as ascidians are needed to learn more about how the molecular evolution of mechanotransduction channels ties into the cellular evolution of hair cells (Fritzsch et al., 2020).

Closer examination of TMC has identified the likely mechanotransduction channel as a specific form of a large family of genes that can be split into two distinct groups in a multicellular organism with only one group playing a role as a mechanotransduction channel in vertebrates. Importantly, this channel can be easily recognized by the highly conserved transmembrane stretches of the protein and thus the evolution of this mechanotransduction protein of vertebrates can be analyzed. Interestingly enough, this family of transmembrane proteins evolved already in single cell ancestors and thus predates its evolutionary late function as a mechanotransduction channel. What exactly this channel protein function is in single cell organisms remains as unclear as the function of the ubiquitously found channel proteins that are part of the mechanotransduction channel. In essence, these findings make it easy to understand the evolution of mechanotransduction as the recruitment of a ubiquitous transmembrane protein into a novel complex and thus a novel function. However, it also raises the problem of the ancestral function of these transmembrane proteins in single celled organisms or other tissue such as osteocytes. At the moment it remains speculative when and how it got recruited to become part of the mechanotransduction (MET) complex of vertebrate hair cells by becoming associated with sensitivity enhancing tethering systems and other not fully understood units of the mechanotransduction complex. Likewise, other putative members of the mechanotransduction complex (*Cib2*) have ancestral roots in protozoans as well as known functions in other tissue. The distribution of *Cib2* and the variable effect of its loss on different mammalian hair cells indicate that reducing evolutionary analysis of such proteins only to their involvement in mechanotransduction will underestimate the pleiotropic functional embedding and consequent selection for different functions in different tissues. Another channel that mediates mechanotransduction in epidermal cells, *Piezo*, is also found in hair cells but may have a distinctly different function and may be associated with mechanotransduction, including recent findings of outer hair cells.

Importantly, mutations in the components that are essential features of the vertebrate MET complex can cause hearing loss, indicating that they are necessary for an as yet not fully disclosed role in this process. Obviously, these major components of a functional mechanotransduction channel need to be present in the vertebrate ancestor as it is unlikely that the mechanosensory hair cell could have evolved **first** in its current form with the channel evolving **after** the hair cell as a morphological entity already existed. Indeed, molecular data suggest that all vertebrates have major components of the mechanotransduction channel based as expected on the ancestry of the hair cells that may represent a modified choanoflagellate, a single cell organism uniformly considered to be the outgroup of animals. It is also noteworthy in the context of the evolution of these channels that mammalian cochlear hair cells that lack one or more of these proteins degenerate thus indicating the essential feature of having the channel ready when vertebrate hair cells began their evolution. One important aspect of the evolution of the mechanosensory channel of the hair cell will be to elucidate when the pore forming protein complex became sensitive to certain ions, acting as a gate that allows rapid diffusion of potassium ions from the potassium rich endolymph into the hair cells upon mechanical stimulation (Fig. 2). Likewise, it needs to be established when in evolution the channel associated with the tip links interconnecting stereocilia of different height started in deuterostome chordates.

Knowing the molecular details of the mechanotransduction complex will allow us to finally unravel how the assembly of this complex might have been shaped before the evolution of the vertebrate hair cell and the cooption of the compounds of the various proteins tethering the channels to other stereocilia (*Cdh23, Pcdh15*) or intracellular actin (*Harmonin*). Conceivably, embedding such channels into a hair cell-like configuration may simply have enhanced the sensitivity of a mechanotransduction channel through its unique association with intra and extracellular tethering molecules, now identified as *Cdh23/Pcdh15* and *Harmonin*. These tip-link connecting proteins are members of the cadherin family that evolved already in single celled organisms. It should be noted that each of the molecules participating in the mechanotransduction complex has or will likely be discovered to have additional functions. For example:

(a) *Pcdh15* mutation may cause severe mental retardation.
(b) *Tmc* channels are also highly expressed in bone where they may function to measure mechanical stress to induce bone plasticity upon load.
(c) Lack of *Cib2* is causing dysfunction of mechanotransduction in auditory hair cells, but *Cib2* is also involved in certain aspects of calcium homeostasis in many cells.

Such known or suspected additional (pleiotropic) functions (Jeong et al., 2022) in other tissue indicate that each of these molecules may have evolved for a different purpose and its function as an integral part of the mechanotransduction machinery of the vertebrate hair cells of the ear may be an example of cooption. Once the molecular details of the functional assembly of the vertebrate hair cell mechanotransduction machinery are clarified, one can begin to unravel the evolutionary details of its constituting parts and reveal how they were assembled and, most importantly, how their assembly correlates with the evolution of the vertebrate mechanosensory hair cell itself. It thus is paramount to understand the evolution of the vertebrate mechanosensory hair cell (Fritzsch et al., 2020) next to begin to integrate the assembly of multiple subunits of the mechanosensory channel complex, that each may have evolved independently, with the evolution of mechanosensory hair cell of vertebrates enabling exceedingly sensitive mechanotransduction.

1.3. New cells are generated by new molecular pathways

Above, I outlined how the evolution of mechanotransduction channels may essentially have followed a molecular transformation of channel proteins that in their basic form are already playing a role as part of a pore across the lipid bilayer of the cell membrane of a single celled ancestor to be opened under osmotic stress, allowing communication between a cells intracellular and extracellular fluid. Obviously, the evolution of the vertebrate mechanotransduction complex is tied in a yet unclear way to the evolution of the vertebrate mechanosensory hair cell, but at least the evolution of the mechanotransduction complex might soon be understood now that core components of the mechanosensory channel protein complex are identified (Fig. 2). The evolution of this vertebrate mechanotransduction cell (the so called 'hair cell') can be broken down into two basic features or modules of cell function and its development:

(a) evolution of the cellular morphology that permits the insertion of the mechanotransduction channel into modified microvilli or 'stereocilia' to improve its sensitivity through intra- and extracellular tethering and transmission of the transduced mechanical stimulus to second order neurons.
(b) evolution of molecular regulation that allows only a set of dedicated cells to develop into mechanosensory cells within a specific organ, the inner ear or the lateral line organs, and guide the development of the mechanosensory transduction morphology, including the insertion of the mechanotransduction machinery, into the cellular processes during cellular maturation as well as synaptic transmitter release.

How evolution may have shaped the development of these cells to endow them with these unique features can only be understood by embedding the problem of mechanosensory cell evolution into the evolution of multicellular organisms out of single celled ancestors. The next set of considerations needs to appreciate the timescale differences: single cell organisms have existed on earth for ~3.5 billion years, whereas multicellular organisms are known for ~750 million years. The absolute time during which evolution could shape the genetic base of cell development and mechanosensory channels is roughly 4 times longer for single celled organisms. Given that the generation cycle is certainly much shorter in single celled organisms compared to most multicellular organisms (days compared to years), the difference in cell cycle length adds orders of magnitude more opportunities to evolutionary change via mutation of the channel proteins and developmental programs of the single celled organism. If single cells were divided about once every day this would allow for ~1 trillion times more replication of DNA. In contrast, if a multicellular organism reproduces just once per year it would provide over 1000 times less opportunity to mutate the DNA (even less if generation cycles are mostly over 1 year as is the case for vertebrates). As with components of the mechanotransduction channels that may have predated the evolution of multicellular organisms, so may the molecular basis of cell fate determination predate the evolution of multicellular organisms. In fact, transcription factors with a basic Helix-loop-Helix (bHLH TF) domain that seem to be at the basis of cellular diversification in metazoans, can act as terminal differentiation factors to experimentally drive one cell to adopt the fate of another cell as originally demonstrated by Weintraub when he converted fibroblasts to muscle fibers by exposing them to the bHLH TF *MyoD*. Importantly, bHLH TF are already found in single cell organisms (Fig. 3) and their multiplication and diversification is certainly an important step in multicellular diversification. Below I will therefore explore this suggestion in more detail in the context of bHLH transcription factors that regulate certain aspects of cell fate in single cells and, after multiplication and diversification, in multicellular organisms.

Comparative molecular data have confirmed what morphological data have long suspected. A small group of unicellular organisms, the choanoflagellates, are the single celled organisms from which all multicellular animals (Metazoans) evolved. These single celled organisms also exist as colonial forms, groupings of individuals that remain independent individuals despite being physically associated. Groups of morphologically identical cells may be attached to a common substrate or are engulfed in a common matrix when free floating. Important here is that choanoflagellates have a clear cellular polarity, whether they are alone or in colonies: one side of the cell, the apex, has a flagellum that propels the animal or food, depending on whether they are attached or not with the opposing part of the cell, the base, providing the attachment. Cells that are attached can use the flagellum to generate a stream of fluid through a ring-shaped collar formed by interconnected microvilli, allowing the organism to filter out food particles that will be ingested.

Defining Novelty in the Neurosensory System

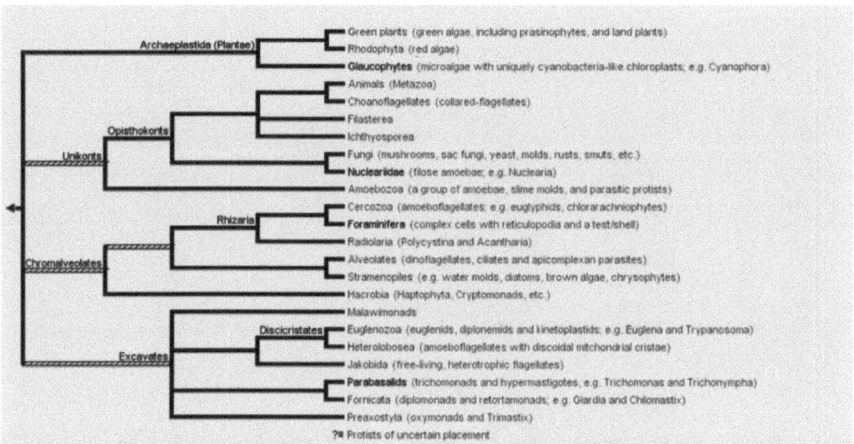

Figure 3: Current understanding of eukaryote interrelationship and derivation of multiple lines of multicellular plants (red algae, green algae, brown algae, land plants from Archaeplastidae), animals (Metazoa), and fungi (both from different lines of Opisthokonts). (Published by (Keeling et al., 2009)).

The most recent molecular evidence currently available supports those sponges are the outgroup of other metazoans, dismissing earlier claims about comb jellies being the sister taxon of all other metazoans. Understanding cellular diversity that leads to the many different cell types of bilaterians needs to start, therefore, by understanding the cell types of sponges and their molecular diversity. Adult sponges have several morphologically distinct cell types: choanocytes (resembling choanoflagellates; Fig. 4), epithelial cells, lining both the internal channels as well as the exterior, possibly several additional cell types in the matrix that makes up most of the sponge matrix, and sensory cells near the ostium. Importantly, sponges have already a-class bHLH TFs that may regulate cell type specification, but experimental evidence is still missing in sponges. Conceivably, the various cells in the matrix are molecularly heterogeneous despite morphological similarities. Importantly, sponges have no morphologically identifiable neurons. Nevertheless, a sponge bHLH gene expressed in developing frog embryos can regulate neuronal differentiation in the epidermis, indicating a high level of conservation of DNA binding sites (so called E-boxes) that allow regulation of ~1000 downstream genes (Wollesen et al., 2015). Some of these genes are also transcription factors regulating in turn expression of hundreds of genes to initiate neuronal cell fate commitment and differentiation. It is this cascading effect that ensures cell fate determination and differentiation during the embryonic development of metazoans most clearly demonstrated in the sequential assignment of cell fate out of a single precursor in the retina. Expression of vertebrate bHLH genes in sponge

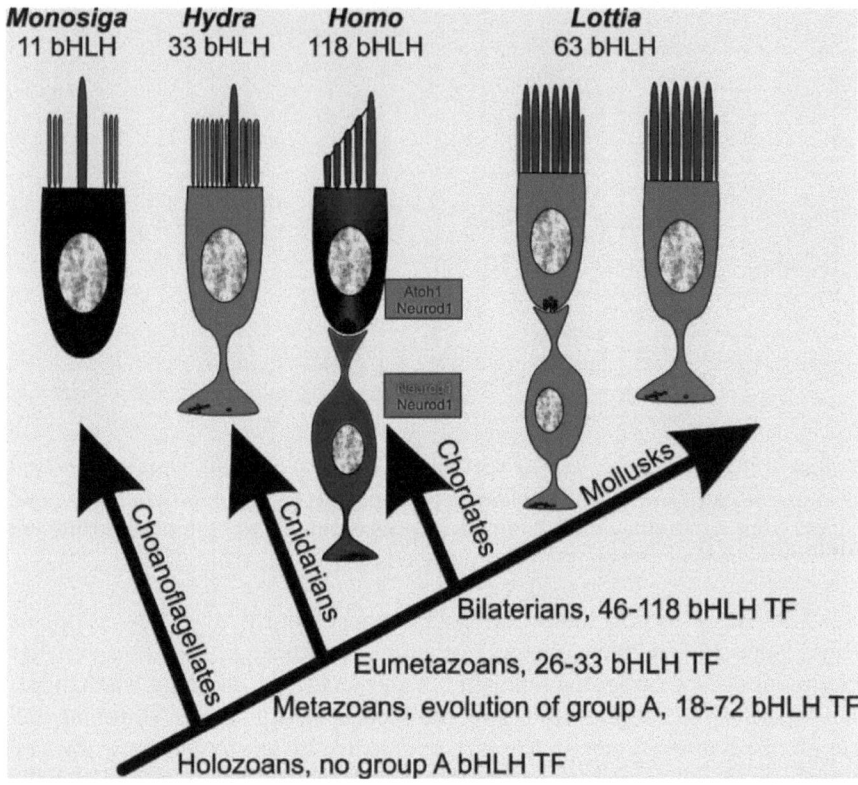

Figure 4: Molecular data show that the evolution of metazoans is accompanied by the multiplication of the bHLH transcription factors from around 10 in single celled ancestors to over 100 in mammals. In particular, the class A bHLH factors, unique to metazoans, multiplied. While details of the roles of bHLH TFs function in single celled organisms have not been experimentally established, it is conceivable that they play a role in switching single cells between a growth phase and a cell division phase. The bHLH TF Myc is a major player in cell proliferation regulation and an important factor in cancer development in vertebrates, with 50% of all tumors showing mutations in Myc-c. Alternating between sexual and non-sexual reproduction and a phase during which the cell growths through feeding could be the initial function of stage specific expression of bHLH TFs. No matter the outcome of this speculation about bHLH function in unicellular organisms, most metazoan cells are undergoing periodic up- and downregulation of expression of different bHLH TFs during development to move through various phases of cell proliferation, determination, and differentiation. This oscillation of gene expression allows precursor populations to multiply while maintaining the commitment to generate even more cells of an already determined general type (for example, neurons versus epithelial cells) while also generating populations of postmitotic cells ready to undergo differentiation under the guidance of different, sequentially expressed transcription factors (Pan et al., 2012).

cells could reveal how much conservation exists across phyla, but the inverse experiment of expressing mammalian bHLH genes in sponges has not been conducted, even though sponges can be easily cultivated. Such experiments could reveal the potential of morphologically highly similar cell types of sponges to respond differently to vertebrate transcription factors. Such experiments, repeating the original experiment of Weintraub to show the transdifferentiation capacity of *MyoD* on fibroblasts to differentiate muscle fibers or of *Neurod1* on fibroblasts to differentiate neurons, could reveal the possibly camouflaged potential of sponges to respond to vertebrate bHLH transcription factors with some degree of cellular differentiation beyond there at best genetic but not morphologically apparent diversity.

In fact, as much as fibroblasts show a hidden differentiation potential that can be revealed by exposing them to certain transcription factors so could the cellular diversity of metazoans be a consequence of prominent expression of a given transcription factor combined with reduced expression of others driving alternative fates. Much like the sensory receptor cell diversity of metazoans reflects the sub-functionalization of multiple incompletely specialized sensory molecules into more specialized metazoan cells, so could multiple cryptic cell morphology potential be isolated through selective transcription factor expression (Fig. 4).

In analogy to the extremely large genome of amoeba, it could well be that the large genetic diversity of sponges with more protein coding genes compared to humans, is associated with too many factors driving cellular differentiation being combined in cells that remain, because of the overlap, morphologically indistinguishable despite potentially rich molecular diversity. Much like free-living cells are generalists in terms of sensory capacity so are fibroblasts, like sponge cells, generalists in terms of cell morphology. However, with proper molecular challenge, the potentially present molecular subgroups could be isolated as only a certain percentage of cells might respond to a given transcription factor challenge. In essence, sub-functionalization and individuation of sensory channels to distinct sensory cells require specialization of generalized fibroblast like or epidermal like cells to become morphologically distinct receptor cells carrying only the selected sensory property: a generalized photo-, mechano-, the chemo-receptive cell has evolved into distinct chemo-, mechano- and photo-receptive sensory cell types (Fig. 3). Below I will outline how far we are in our understanding of this molecular cellular process that leads to isolation of highly sensitive mechanotransduction to morphologically uniquely adapted sensory cells that enhance the molecular process of the channel through tethering to extra and intracellular connectors.

One basic aspect underlying morphological cellular diversity is the multiplication followed by diversification of transcription factors that drive this process. Among those genes, basic Helix-Loop-Helix transcription factor coding genes (bHLH TFs) serve a special role as they tie into two principle cellular processes: cell proliferation and cell diversification, both essential features of metazoans. Combined with molecular mechanisms that ensured proper sorting of chromosomes such as the guanylate cyclase (gk) complex this ensured the coordinated diversification of multicellular specification within an ordered assembly of cells able to form epithelia with distinct functions. Overall, multicellularity evolved multiple times but led to large bodies with diversified cell types only in the above outlined 6 cases. Beyond what has been outlined above, multicellularity, as far as it can be observed within a group, seems to follow a progression from unicellular to colonial to multicellular transformation. Obviously, changes in gene expression regulation through modifications of the gene regulatory network (GRN, (Peter and Davidson, 2015)). Such GRNs can produce what has been dubbed as dynamical patterning networks (DPMs) that involve one or more GRNs and physical properties to drive development and evolution (Newman and Bhat, 2009). The above outlined bHLH transcription factor belongs to these highly conserved GRNs that function within eukaryotes, including all multicellular forms (Niklas, 2014). Below I will explore the role of bHLH transcription factors in the evolution and development of neurosensory systems, a unique feature of animals not shared with the other five multicellular organisms that evolved out of opisthokonts. It remains to be seen if this indicates a chance event or how much environment plaid a role that certainly was unique for metazoans compared to other multicellular organisms.

A simplified version of a hypothetical cell that indicates some of the extracellular signals that affect the regulation of cell fate decisions to proliferate or to differentiate, by balancing the signals of pro-proliferative bHLH TFs such as *Myc, ID, Hes,* and pro-differentiation bHLH TFs such as *Atoh1, Neurod1, Neurog1* (Figs. 5, 6). When this complicated interactive network of various proteins evolved in its rudimentary form is unclear, but it was already needed to regulate in single cells the transition from one state to another, such as from feeding to dividing (Fritzsch et al., 2015). Obviously, many extracellular signals can regulate the expression of the intracellular mediators in metazoans to ensure that proliferation and differentiation are happening in the proper cellular context. For example, alterations in the diffusible molecule *Shh* can affect the level of *Myc* expression accelerating or decelerating cell cycle progression, enabling rapid tumor growth such as medulloblastoma, in the developing cerebellum the long-lasting proliferation of external granule cells that generate the largest number of neurons in the brain of mammals and some bony fish is regulated by Shh signaling via its receptor Smoothened (*Smo*; Fig. 5). Expressing a constitutively active, constantly signaling Smo protein suffices to induce tumor transformation in the cerebellum and cochlear

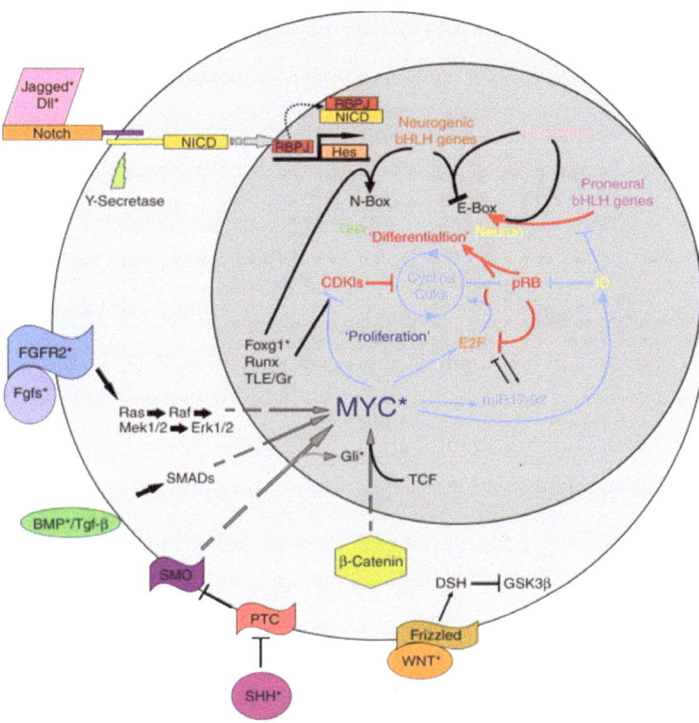

Figure 5: Myc illustrates the complex, interrelated, and essential balance between proliferation and differentiation. In the presence of nearby neurons, Jag1 binds to Notch1, and subsequently, gamma-secretase is free to cleave the notch intracellular domain (NICD). NICD translocates to the nucleus to form a complex with RBPJ to positively regulate the transcription of the neurogenic *Hes/Hey* family of bHLH transcription factors. Hes/Hes in combination with E-proteins inhibits binding to the E-box and promotes binding to the N-Box, resulting in shifting the balance from differentiation of neurons to differentiation of glial cells within the system. In the absence of delta/notch signaling and *Hes/Hey* transcription factors, proneural bHLH transcription factors bind to E-proteins to promote the formations of neurons. Hence, there is a balance between neurons and glial cells in the face of differentiation. However, in the system, there exists a balance not only between glial cell fate and neuronal cell fate but also between proliferation and differentiation. Retinoblastoma (Rb) in its unphosphorylated state binds the E2F family members. When Rb is bound to E2F, cells cannot pass the G1/S checkpoint, exit the cell cycle, and undergo differentiation into either neurons or glial cells as described above. However, both cyclins and the inhibitors of differentiation (IDs) can free E2F from Rb, allowing E2F to bind to S-phase-promoting genes and allowing the cell continue to DNA replication. ID has an additional function that promotes proliferation; ID can directly bind proneural bHLH transcription factors and inhibit their binding to the promoter regions of target genes. This inhibits differentiation and promotes proliferation, illustrating the tight regulation and

(*Contd.*)

Figure 5: (*Contd.*)

importance of the balance between proliferation and differentiation. On the other hand, cyclin-dependent kinase inhibitors (CDKIs) inhibit the cyclin-dependent kinases (CDKs), resulting in net proliferation. Upstream of the IDs, cyclins, E2Fs, and CDKIs is *Myc*. The proto-oncogene *Myc* is an important node in this pathway, as illustrated by the many developmental defects and tumorigenetic defects caused by *Myc* deregulation. *Myc* is regulated by several pathways, including the *Wnt/β-catenin*, *TGF-β* or *BMP/SMAD*, *Fgf/Erk1-2*, and the *Shh/Smo* pathways. In short, the *Fgf* ligands diffuse to their receptors which bind the *Fgf*'s at the surface of the cell and signal through second messengers, the RAS intracellular pathway to influence *Myc* signaling. The *Bmp*'s signal through the *SMAD* pathway intracellularly acts indirectly on *Myc* function. *Shh*, which binds with Patched and interacts with Smoothened at the cell surface influences *Myc*. Lastly, β-catenin constitutively acts in the presence of *Wnt* through disinhibition of β-catenin breakdown. Together, *Myc* is an integral node integrating both upstream and downstream pathways. Furthermore, the downstream regulation by *Myc* shows the highly complex, interconnected, and essential balance between proliferation and differentiation. The larger circle represents the cell as a whole, while the smaller circle is the nucleus. Pointed arrows indicate a favored pathway, while blunted lines represent inhibited pathways. In general, blue colored lines indicated steps that favor proliferation, while red colored lines indicated steps that favor differentiation.
From (Kopecky and Fritzsch, 2011).

nuclei, causing medulloblastoma (Jahan et al., 2021). This rapidly growing tumor that constitutes over 30% of child brain tumors is investigated to find means to eradicate it or at least slow it down. Blocking Smo signaling using small molecules such as cyclopamine or vismodigib counteracts tumor formation and proliferation and vismodigib is clinically the current drug of choice to slow down medulloblastoma growth. Another bHLH TF, *Atoh1*, is also needed to keep the proliferation of medulloblastoma growing and loss of *Atoh1* signaling can counteract medulloblastoma formation but also eradicates the formation of the most numerous neurons in the mammalian brain, the granule cells of the cerebellum. I will come back to this function of *Atoh1* once I have introduced the function of *Atoh1* in developing hair cells.

Generating the diversity of morphologically identifiable metazoan cell types to generate multiple tissues is not only directly proportional to the multiplication of bHLH transcription factors and their interactions but requires additional modification of the general signaling pathway, leading to cell type specific differentiation. Nevertheless, it is important to note that the number of bHLH genes increased 10-fold from choanoflagellates to mammals (Fig. 4) whereas the total number of genes seems to have just tripled (from ~9000 to ~24000). However, there is an approximately 100-fold increase in total DNA from Choanoflagellates (~50 megabases) to Humans (3.235 megabases; haploid). Among the alterations now identified among bHLH genes are positive and negative feed-forward and feedback loops as well as micro RNAs (miRs). miRs can serve as a second filter to sharpen the expression of mRNA species in a given cell and seem to have evolved at the root of bilaterians metazoans that have many more morphologically distinguishable cell types compared to sponges, cnidarians, placozoans, and comb jellies. Lack of miR leads to a dramatic loss of neurons and sensory cells whereas expression of certain miRs can initiate cellular differentiation. Most importantly, expanded sets of miRs in salamanders may mediate the extraordinary regenerative capacity of these animals, enabling cells to dedifferentiate, proliferate and eventually differentiate again to build a new limb out of a blastema of undifferentiated cells that form after amputation.

More recently, it was also discovered that not only is the ~2% of known protein coding DNA transcribed into mRNA, but over 60% of DNA is transcribed into long noncoding RNA (lncRNA). While details are not fully worked out, lncRNA cooperates and/or antagonizes certain miR species, thus causing additional cell type specification alterations and fine tuning of cell types. In addition, many vertebrate genes are broken down into multiple exons separated by introns. Such a gene organization allows to form different mRNA transcripts through differential splicing of some but not other exons into a given, cell type specific mRNA. Proteins that regulate that process can determine which of several possible transcripts and thus proteins will form in a given cell, adding another layer of cell fate determination control that we do not yet fully understand.

Combined, all these additional processes need to be evaluated experimentally before we can begin to grasp how nature generates and selects cellular diversity in evolution and how much development of cell type diversification recapitulates some or no aspect of cell type diversification evolution. I will explain this complex regulatory scheme of cell types of specification further after I have described the development of the mammalian hearing organ and its cellular constituents in a later chapter. For the moment, it suffices to understand that different stages of a cell cycle and the different stages of progression toward differentiation require each cell type specific sequential activation of many genes. Collectively this activation ensures that a given step of development is properly executed to safeguard that the next developmental phase can be securely implemented and bHLH TFs play a central role in several steps of this cell fate process.

1.4. Understanding metazoan evolution

Understanding metazoan evolution will ultimately hinge on understanding how this cell cycle/cell differentiation cascade evolved to multiply precursors. Metazoan evolution depends on cell cycle and differentiation:

(a) to generate more cells from a single fertilized egg to make the ~1 trillion cells of the human body and
(b) differentiate them into over 100 distinct cell types for a specific tissue.

Beyond the function of bHLH TFs in the context of proliferation/differentiation, one needs to understand how the above outlined general mechanism specifically ties into the apical/basal differences of cells such as choanoflagellates and vertebrate hair cells (Figs. 4, 5). Only the apex of these cells develops microvilli and kinocilia. How differentiation of microvilli filled with polymerized actin is specifically driven in apical and not basal regions, how the kinocilium is developed only between the forming microvilli and, unique to vertebrate hair cells, moved into an asymmetric position relative to microvilli that will eventually transform into 'stereocilia' are only fragmentally understood. What is clear is that these apical protrusions of vertebrate hair cells are foremost composed of a polymerized actin core stabilized by small GTPases that regulate the polymerization and interactions with other complex proteins. Many of these proteins have recently been identified in the choanocytes, the choanoflagellates-like cells in the chambers of sponges that circulate water through the sponge and filter detritus for feeding the animal. Unfortunately, it is unclear if major components of the mechanotransduction complex, clearly identifiable as highly modified genes with limited sequence similarity in sponges, are expressed in sponge stereocilia.

With this caveat of uncertainty in mind, it is interesting that morphological series of alterations in these apical specializations of metazoans exist that can be interpreted as a gradual change from a choanoflagellate/choanocyte like organization in a unicellular and basal multicellular organism to the highly asymmetric organization of stereocilia and kinocilium in vertebrate hair cells (Fig. 6).

The series of morphological changes of the cellular apex and the changes in mouse hair cells in various mutants suggest the following overall scenario: Choanoflagellates and choanocytes of sponges present the basal features of a motile kinocilium surrounded by interconnected microvilli that function to filter food particles. Whether or not they also have a sensory function is unclear but could be investigated now that candidate genes for mechanotransduction channels have been identified. A separate line generated ctenophores (Burkhardt, 2022; Fritzsch et al., 2015). Interestingly enough, epidermal cells with two immotile kinocilia seem to play a sensory role and keep track of water flow in sponges. Via still unclear mechanisms they can stop the beating of the choanocytes and cause contraction of the sponge to expel detritus. It is important that these are kinocilia without associated microvilli that are more akin to the immobile kinocilia specialized for mechanosensation found in mollusks and insects. This differs from vertebrate hair cells where kinocilia do not play a role in mechanosensation and can be lost in the mammalian hearing

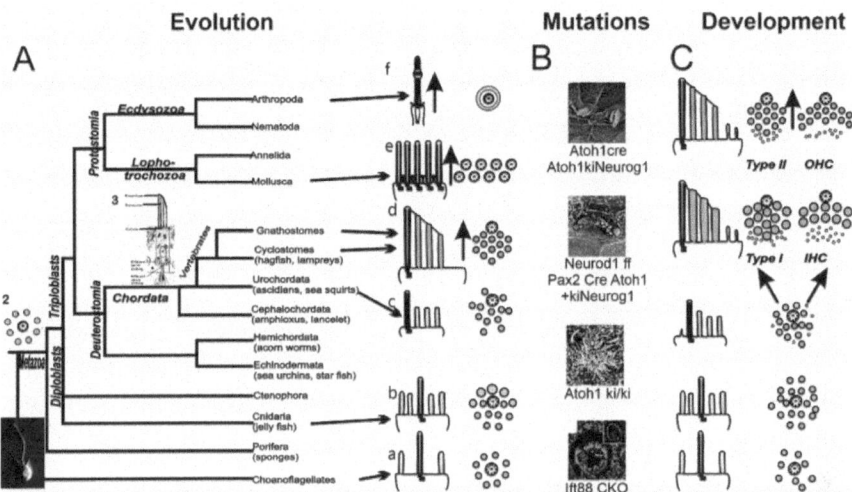

Figure 6: Evolution and development of the apex from centric to acentric organization. Evolution (A): The kinocilia shifts from a centric position (2) surrounded by microvilli/stereocilia found in choanoflagellates and choanocytes of sponges to an acentric position (a-d) in Deuterostomia, establishing cell polarity and allowing directional stimulation in vertebrates (3). Protostomia show divergent specializations for mechanosensation (e, f). Movement direction is indicated by the arrow. Mutation (B): Various mutations that affect the stereocilia. *Atoh1-cre, Atoh1kiNeurog1, Neurod1f/f Pax2-cre, Atoh1+kiNeurog1*, and *Atoh1ki/ki* mice develop stereocilia aberrations suggesting that planar cell polarity is influenced by *Atoh1* levels. *Itf88* CKO mice can develop circular stereocilia in the absence of kinocilia. Development (C): Hair cells develop first into a central kinocilium surrounded by microvilli followed by a shift of the kinocilium into an acentric position and alteration of microvilli diameters and height to form stereocilia (bottom to top). Only the organ of Corti hair cells loses secondarily the kinocilium (top). (Fritzsch and Elliott, 2017).

organ. In vertebrates, only highly modified microvilli, the 'stereocilia' hold the mechanotransduction complex. These data suggest that mechanotransduction is morphologically distinct, likely depending on different channel proteins whereas certain aspects of molecular control of development rely on the same, interchangeable transcription factor. Since neither the history of the acquisition of mechanotransduction complexes nor its regulation by transcription factors during development is well understood, I will not explore this critically important aspect of cellular functionalization further.

Morphologically similar cells with microvilli and a central kinocilium are found widespread among basic metazoans whereas sensory cells with two kinocilia are only known in addition to sponges for the organs of de Quatrefages of lancelets that have some molecular similarity to mechanosensors of other animals such as expression of specific miRs.

For example, the miR-183 complex is highly conserved among bilaterians and seems to be associated with mechano- and/or photosensitive cells across phyla. Overall, these data suggest that the transformation of motile kinocilia to sensory kinocilia may be as old as metazoans, but it remains unclear how much of the molecular machinery of the vertebrate mechanosensory complex exists already in choanoflagellates, diploblasts and basal deuterostomes.

Beyond the molecular and cellular basis of mechanosensation, the sensory input guided behavior of metazoans may predate the formation of a nervous system with neurons. Such communication from sensory cells in sponges seems to rely on the excretion of molecules using ancient vesicular release systems already evolved in single cell organisms and used, for example, in the communication of slime molds that can form transient, multicellular aggregates for propagation. These data suggest that it is not only molecules involved in mechanotransduction and vesicular release that are conserved across phyla but also organelles dedicated to aggregating the mechanotransduction machinery. Such aggregation to kinocilia or stereocilia is conceivably based on the topological organization of these cellular protrusions because such channel proteins can connect to both extra- and intracellular matrices within and between stereocilia, boosting mechanosensitive as diagrammatically outlined in Fig. 4.

While this general trend is clear, the details of both mechanotransduction channels and microvilli or kinocilia distribution across metazoans as well as the details of membrane insertion as well as intra- and extracellular connections seems to vary across phyla. Nevertheless, only a limited number of channels and their connection molecules have been identified across phyla, and many share at least some features. As outlined in Fig. 5, the vertebrate mechanosensory channel and possibly its connection to the intra- and extracellular matrix seems to be more conserved in diploblasts and deuterostomes whereas Protostomia has evolved rather different channels and anatomical features for mechanotransduction embedded with a kinocilium as the morphological manifestation of mechanotransduction. More molecular analysis across phyla is needed to reveal how the single celled ancestor's mechanosensory system was organized and how it was possibly rearranged in metazoans to be differentially distributed, possible via multiplication and diversification of several channels, to function in structurally different apical stereocilia/kinocilia arrangements (Figs. 5, 6).

Among deuterostomes, current evidence suggests that the formation of vertebrate hair cell-like cells predates vertebrates: The distinct asymmetric organization of the kinocilium and organ-pipe organization of stereocilia is only found in mechanosensory hair cells of vertebrates whereas non-vertebrates have variably asymmetric arrangements of microvilli around a more or less central kinocilium. From these morphological differences, it follows that one need not only understand how much of the mechanotransduction channel and its connections to intra- and extracellular matrix is conserved (Fig. 4), but also when the molecular mechanisms evolved for moving the kinocilium from a central position, found in basal metazoans and single celled ancestors, to the eccentric position of the adult hair cells of craniates. This evolutionary change is achieved through altering a developmental program that starts the vertebrate hair cell development resembling that of choanocytes, with a central, apical kinocilium surrounded by microvilli (Fig. 6). What happens during development is that one side of a developing hair cell apex absorbs microvilli whereas, the other side transforms microvilli into the larger and more actin containing stereocilia. While the microvilli on one side are resorbed, the kinocilium is shifted toward the area devoid of microvilli. Some aspects of this molecular interaction during development that defines the apical polarity, reduce the microvilli and move the kinocilium have been unraveled but as of this writing, it appears that not all components are known nor is their interaction entirely validated by critical experiments. What is clear is that many mutations can cause a developmental arrest of this process leading to unlikely hair cell apex morphologies (Fig. 6).

For example, either the lack of the bHLH transcription factor *Atoh1*, its substitution by another bHLH transcription factor (*Neurog1*), or the loss of all micro-RNA species can result in a hair cell apex characterized by a central kinocilium surrounded by multiple microvilli, always in a slightly different appearance (Jahan et al., 2015b). Obviously, these mutant phenotypes are not a simple recapitulation of an evolutionary more basal hair cell phenotype (Fig. 5). Instead, they show that basic principles of evolutionary changes from a central to an acentric kinocilium require a sophisticated molecular cascade that can be disrupted at different levels through the elimination of different genes within a development regulation core transcription factor complex, leading to a variably arrested hair cell apex phenotype.

In keeping with the overall quest underlying this writing, future work will need to find out how the insertion of mechanotransduction complex units into stereocilia correlates mechanistically with the asymmetric development of hair cells. Further, how the various mutations that affect stereocilia elongation, stereocilia diameter, stereocilia distribution, stereocilia interconnection, and stereocilia actin bundling touches insertion of the various components into the stereocilia tip as well as their assembly requires also more work. Armed with the knowledge of at least major components needed for a functional mechanotransduction complex gained over the last 10 years, one can now begin to analyze how the hair cell fate determining molecular cascade generating the asymmetric hair cell apex ties into the development of the apical specialization by regulating the expression of factors essential for the reorganization of the microvilli into stereocilia, the movement of the kinocilium and the upregulation and insertion of the mechanotransduction complex. In particular, genetic manipulations beyond simply knocking out a suspected gene are needed to study not only loss of function effects (what can the system still do if I cut off one leg) but also gain of function manipulations (what can a system do with a third or a fourth leg).

Some of these manipulations are now possible through the quantitative regulation of a single crucial transcription factor in hair cell development, the bHLH gene *Atoh1*, or by replacing *Atoh1* with closely related other bHLH genes such as *Neurog1* that can activate some, but not all of the downstream effectors of *Atoh1*. Given the emerging complexity of intracellular interactions, in part driven by as yet incompletely characterized feedback and feedforward loops of these transcription factors, but also in part related to the evolving transcription factor binding properties and their likely interactions within a cell pushes such an analysis beyond the current technical abilities. I will pick this topic up again after I explore in more detail the cascade of transcription factors needed to transform a naïve epidermal cell into a prosensory otocyst precursor cell ready to start the differentiation as a viable hair cell.

Peculiar to hair cells associated with the mammalian hearing organ, the organ of Corti is a late developmental loss of the kinocilium. Given the emerging role of kinocilia in intraepithelial signaling and its integration into the asymmetric development of the hair cell apex, a kinocilium is apparently essential for any hair cell development but must be eliminated for normal function in the mammalian hearing organ that will not work with a kinocilium. I will explain that argument in more detail below once I have introduced how the mammalian organ of Corti functions differently from vestibular sensory organs. For now, it suffices to realize that the evolution of the mammalian hearing organ acquired a unique developmental addition to the vestibular hair cell program that led to the postnatal loss of the kinocilium after its use to integrate diffusible signals for synchronized polarity development of multiple hair cells across a given sensory epithelium has been accomplished. Beyond a description of this process of late embryonic, early postnatal loss of kinocilia, nearly nothing is known about how this comes about mechanistically nor if, and possibly how, this loss correlates with the auditory hair cell expression Cib2 that is crucial for mechanotransduction and viability. Such information is essential should we ever be able to overcome current roadblocks of cellular hearing restoration that has thus far only managed to generate vestibular-like hair cells with yet unclear long-term viability. I will explore this aspect in more detail below.

In summary, this step-by-step outline describes how a single cell can evolve into a sensory cell of a multicellular organism through the functional reorganization of existing transcription factors. Multiplying transcription factors allowed to regulate interactively the proliferation of many cells out of a single cell as well as the differentiation of these cells into multiple cell types and tying them into regulating the evolving apical mechanotransduction complex. Importantly, critical incompletely understood aspects of this process are highlighted to guide future research that can clarify the details of these processes and define how developmental modules of the ancestral single celled organism were modified to generate the apical and basal cell specializations of a vertebrate hair cell.

Multicellularity evolved at least three times independently: in plants, fungi, and animals. Interestingly, only animals used existing unicellular sensory capacity to evolve dedicated sensory cells that aggregated into sensory organs. Neither plants nor fungi have evolved such specialized sensory cells or organs and also lack the formation of a nervous system that can integrate stimuli from different sensory cells or organs into a cohesive organismal response. Therefore, while single cells ancestral to all three multicellular lineages can sense gravity, chemicals, and light, only animals build on the existing molecular repertoire by making cells in which molecules enabling a specific sensation were accumulated. Plants and fungi have not evolved beyond the single cell sensory capacity, suggesting that specific aspects of animal multicellularity drive neurosensory cell formation. To achieve this, animals build on specific sets of transcription factors to define and differentiate specific classes of cells as an adaptation to their lifestyle. Only animals with sophisticated sensory inputs have evolved neurons that allow information processing to control directed motor output. Once animal ancestors had achieved segregation of sensory molecular distribution into distinct sensory cells for information gathering, an integrating nervous system becomes essential to guide proper responses of the whole organism. Aiming to understand neuronal evolution without understanding sensory cell and organ evolution is therefore meaningless as neurons evolved as a consequence of sensory molecules for stimuli segregated to discrete cells to maintain a coordinated organismic response that was so readily available at the single cell level.

1.5. Defining hierarchy of events to understand cascading effects of changes of neurosensory evolution

Early metazoans, much like sponges, likely had no specialized sensory cells or a central nervous system to integrate sensory information by cells dedicated to receiving a given stimulus. Assuming sponges are not secondarily simplified as their current taxonomic position implies, it appears that sponges show rudimentary neurosensory functions whereby a given stimulus leads to a whole animal response. As far as specialized sensory cells are concerned, it appears that the sensory input for the 'couching response' of the entire sponge comes from a transformed choanocyte-like cell monitoring water intake and extrusion. This transformation of these cells eliminated the microvilli, doubled the kinocilium, and enabled such sensory cells to communicate with the entire organism to elicit an appropriate organismal response. As it turns out, sponges have some aspects of the molecular mechanisms for vesicular release typically encountered on presynaptic terminals as well as certain molecular features usually associated with the postsynaptic specialization: molecularly speaking, sponges seem to have much of the repertoire needed to release synaptic vesicles. Sponges also have molecular features that allow changing the resting potential of cells such as ion-selective channels leading to potassium, sodium, and calcium flow across membranes. In fact, all of these features were inherited from single celled organisms, some of which can communicate via the vesicular release of molecules to alternate the behavior between a free-living unicellular form and a transient aggregated form for reproduction such as fruiting bodies in slime molds. Sponges also have various cadherins that allow them to connect cells to form a multicellular organism.

Many of these molecular features of sponges that enable some degree of cellular communication, cellular adhesion, as well as response to such signals, predate multicellularity and certain formation of specialized nerve cells dedicated to conducting rapid information gathered by more or less specialized sensory cells to effector organs such as glands or muscle fibers. It is noteworthy that among the three lines of multicellularity (fungi, plants, animals) only animals evolved the fast-conducting nervous system that allows sensory information dedicated to perceiving specific aspects of the vast array of environmental signals to elicit appropriate responses within the limits of the motor capacity of a given animal. The above outline evidence suggests that the evolution of mechanosensory channel components occurred before they became integrated into the evolution of apical cellular processes that allowed making the connection of this channel (s) to enhance their sensitivity. Likewise, vesicular release evolved first, before it was coopted into a synaptic complex that enhanced vesicular release for more rapid and sustained cellular communication in a synapse between cells. Similarly, ion channels to conduct action potentials evolved prior to nerve cells and were coopted into nerve cells. How cell fate determining factors such as the above outlined bHLH TFs were not only tied into the evolution of mechanosensation of the apex but also into the evolution of synaptic release mechanisms concentrated at the base remains to be seen and the expression and insertion of ion conducting channels remains to be revealed. The apparently molecularly ancient bHLH TFs such as *Atoh1* play not only a role in organizing the apical specializations of mechanosensory cells but also in regulating synapse formation not only in hair cells but also in the majority of neurons in the central nervous system (CNS) possibly including their action potential conduction mechanisms. In fact, neurons are highly specialized secretory cells interconnected with elongated processes for fast information conduction and communication via highly modified cell junctions, referred to as synapses.

Organisms such as sponges or placozoans with limited ability to escape a predator have little use for a sophisticated nervous system that ca integrate multifaceted sensory stimulation to guide complex evasive motor responses, a response these animals are unable to do so. The evolution of a nervous system is logically tied to the sophistication of the sensory input as well as the possible motor output. For obvious reasons, neither the evolution of a sophisticated sensory system without a motor ability to translate such information into action nor the evolution of a sophisticated motor system without a sensory input to put it to good use seems to be compatible with a selective advantage that provides a higher reproductive success. Simply speaking, animals that can sense an approaching predator but cannot escape or animals that could escape but cannot sense an approaching predator are equally doomed. Simple reflex arcs from a sensor to an effector need to form first and those simple response systems will eventually become more sophisticated sensory-motor interfaces by adding more and more interneurons. Consistent with this scenario is that multicellular plants and fungi have no ability to move and lack a nervous system and sophisticated sensory organs whereas their single celled precursors respond to various stimuli much like single celled ancestors of animals. I will discuss how such changes in sensory-motor interactions can arrive through molecular evolution and can be stabilized by selection in a later chapter dealing with the evolution of the central nervous system. Here it suffices to understand that only in the evolution of animals have the various sensory and motor capabilities found in most single celled organisms ultimately segregated into discrete cells and organs that allow a more refined sensory perception of the environment with dedicated sensory cells and organs, a more sophisticated information processing of this information using nervous tissue and more effective translation into action using specialized muscles and glands.

While this plausible scenario accommodates much of our current understanding, it falls short of taking into account recent molecular and systematic changes in comb jellies. For over 100 years, comb jellies have been combined with jellyfish as coelenterates, but recent data have questioned the taxonomic position of comb jellies. In fact, comb jellies appear in many respects to be the outgroup of all other animals. If so, the presence of a unique gravistatic sensory organ on the apex of comb jellies and neuronal connections of the apical organ to the comb rows that provide the motor output indicates one of two possible scenarios:

(a) Neurosensory evolution happened twice in animals, once in comb jellies and a second independent time in jellyfish and related coelenterates.
(b) Alternatively, neurosensory evolution happened only once and the absence of such specialized cells in sponges and placozoans reflects a secondary loss.

It is noteworthy that among other multicellular organisms such as fungi and plants in which no nervous system ever evolved, the first assumption outlined above would be that such specializations evolved twice and independently in animals. While possible, it appears that the alternative scenario that requires a single gain of a neurosensory system and two losses is equally likely. More details are needed on the overall and detailed similarities of neurosensory development of comb jellies and jellyfish to decide between these competing interpretations.

This increased complexity of sensory input relationship to motor output can best be understood considering simple aspects of sensory-motor interfaces. An example of a simple reflex arc is the monosynaptic knee jerk reflex of mammals. The input of a stretch receptor signal conducted via sensory neuron processes from the periphery directly reaches a spinal motor neuron with the sensory axons forming synapses on those motor neurons. Once a motor neuron is stimulated by the incoming signal, it activates through its synapse on muscle fibers the contraction of the stretched muscle to counteract the stretching. If integrated along an entire organism such actions will result in the contraction of all muscles on the stimulated side of the organism, resulting in a bending movement away from the stimulation source, if at the same time, the muscles on the other body side are relaxed. Such unsophisticated responses could possibly be sufficient for an organism with limited mobility to avoid being eaten by a nearly equally mobile predator, offering a selective advantage. Available fossil evidence of early metazoan biota strongly implies that most were indeed sessile with limited filter feeding capacity. Simple speaking, among the immobile life forms even the slowest mover is the fastest!

In contrast to nearly immobile or sessile organisms, free-swimming organisms are confronted with different challenges requiring some degree of orientation in the water. Of course, one solution is to 'go with the flow' and just drift wherever water currents bring the organism. Given that food availability is different in different layers of the ocean and at different times of the day, animals that have at least some minor capacities to navigate within the water column would have a selective advantage over those that are randomly floating around. Consistent with this suggestion of selective advantage of some capacity of orientation is the distribution of complex, multicellular sensory organs to orient an animal mainly in gravity are only found in free-swimming diploblasts such as comb jellies and jellyfish. In contrast, diploblasts with limited or no mobility (corals, placozoans, sponges) lack such complex sensory organs but have single sensory cells or small aggregates of sensory cells able to monitor touch or flow of water across them and can respond to certain changes in light intensity and chemical composition of the water. It is important to realize in this context that free-swimming sponge larvae seemingly have a more sophisticated sensory response system with little to no morphological specialization in terms of anatomically recognizable sensory cells and organs. In fact, this correlation has been invoked in the past to suggest that the evolution of free-swimming larvae into adults might be at the root of metazoans with a sophisticated nervous system. Be this as it may, the obvious correlation between free swimming versus a sessile lifestyle with more sophisticated sensory organs and a nervous system is undeniable.

This rule of increased neurosensory capacity associated with motile forms of a given organism, or closely related organisms such as cnidarians can be expanded to deuterostomes. Many tunicates have sessile forms with a more limited neurosensory system compared to free-swimming larvae. Those with more developed neurosensory systems as adults are adult larvae, hence their systematic name 'larvaceae.' This distribution of complex neurosensory systems nearly exclusively in free swimming diploblasts has prompted arguments that central nervous system evolution is tied to the need to survive as floating, potential prey, or to orient to the floating food source as a free-floating predator. While this is overall agreed upon, the details are argued about, in particular, the chicken and egg question—if the evolution of a motor system predates that of a sensory system, or vice versa, or if there was an intricate co-evolution of both. I posit here and will explain further below that sensory evolution predates motor evolution. This is particularly obvious in the case of passive and active electroreception, with active electroreception being restricted to a few species able to modify muscle fiber signal output to reduce contraction while maintaining electric signal production. In contrast, the more widespread passive electroreception does not require muscle fiber modification as all muscle fibers elicit electric signals when they initiate contractions.

I will therefore concentrate here only on the sensory input evolution, expanding the discussion beyond the transformation of the above outlined mechanosensory ability of individual cells to several senses [for example, mechano- and photo- (light) sensing] as needed to make my points. The premise I like to develop is that information provided by each one of these sensory modalities needs to be directed to different parts of an emerging central nervous system for more sophisticated responses to process either each of these inputs alone or to generate an integrated composite signal. Achieving this requires the evolution, minimally of sensory cells and possibly sensory organs dedicated to only one sensation to segregate molecular precursors found in single celled organisms to enable mixed photic- and mechanical-sensations. In essence, metazoan sensory cell formation can be regarded as a subfunctionalization of sensory cells to become adapted to perceive a given, more specific stimulus while suppressing all other sensory modalities. Instead of being a 'jack-of-all-traits,' able to sense many stimuli with little sophistication, dedicated sensory cells perceive only a given sensory stimulus, excluding all other sensory input by parsing out and enhancing all input abilities related to this one sensory stimulus.

As already introduced, evolving distinct sensory cell types depends on duplication of cell fate specifying transcription factors that allow allocation of distinct sensory molecules to distinct cell types instead. This process may already have molecularly started in single celled ancestors with organizing the differential distribution and regulating the various molecules involved in mechano-, photo- or chemo-sensation to different parts of the cell. For the following, I will concentrate only on two sensory systems (photo- and mechanosensation) for which we have reasonable molecular evidence that they are derived from a common ancestor carrying both in the same cell through cellular diversification. This diversification was made possible by the duplication of transcription factors driving downstream genes regulating molecular entities enabling one or the other sensation followed by selective expression of the functional components in different cell types, enabling distinct sensory cells to be stimulated by either light or mechanical energy.

Like mechanosensitive channels that can channel the changes in their gating properties upon exposure to light can be traced back to single celled organisms. Essentially, some single celled organisms respond to either mechanical or photic stimulation, or both, by changing the direction of swimming, accelerating or stopping swimming. Indeed, even in single celled fungi, obvious light sensitive organelles exist in a specific part of the cell and this is also true for single celled plant precursors. With the appearance of multicellular animals, these sensory capacities were eventually segregated into distinct cell types that integrated into distinct organs. Sponges have no complicated sensory organs, indicating that animals in the specific niche occupied by sponges may have parts of the molecular disposition to generate complex sensory cells and even organs but have never executed that potential. In contrast to sponges and placozoans, some jellyfish have complicated lens eyes, comparable to those of vertebrates or cephalopods, which respond to light using retinal pigment that is evolutionarily related to that of vertebrates. In contrast, comb jellies have no morphologically distinct eyes or photoreceptors but have general light sensitivity. In addition, both jellyfish and comb jellies have morphologically distinct statocysts to perceive the gravity mediated mechanical stimuli comparable in function to the gravity receptor part of the vertebrate ear. The cellular and organ differences between comb jelly, jellyfish, and vertebrate statocysts/ears suggest they evolved independently. However, the morphological differences of the adult sensory organ do not exclude those identical molecules that guide certain steps of development as explained in more detail next.

Combined, eyes and statocysts allow jellyfish to orient in water either to swim deeper toward gravity and away from light or toward the light when they are in deep dive, by swimming away from gravity towards the surface of the water column. Obviously, the distinct and antagonistic input of light and gravity into the swimming behavior can only be accomplished once the two sensory systems driving those responses are segregated into different cells that will be differentially distributed and eventually aggregated to form different sensory organs. Since light and gravity will never point in the same direction, segregating both inputs allowed for a more sophisticated orientation response beyond just avoiding or being attracted to either or both the gravity of the earth and the light of the sun/moon. Jellyfish show that the evolution of such complex sensory organs as eyes, which baffled the likes of Charles Darwin, seemingly can occur even if the organism having such complex organs has just a nerve net to process the information driving swimming behavior only sufficient to change slightly directions to move up and down in the water column. Since these organs are strategically placed, they will directly affect the contractility of their local area without a need for sophisticated integration. In essence, integration is in the hardware of the sensory organ distribution and the activation of local contractile elements requires little computing power to extract and relate the relevant signal to control swimming. This is comparable to a lens eye using the physical properties to automatically generate such images, whereas a simple light sensing array without the lens and pinhole can only form a complex image using sophisticated computer programs. Selective ablation of one or more organs could demonstrate whether or not the organism as a whole can compensate. The prediction is that it will not be able to do so as orientation is part of the hardware. However, it should be noted that modern jellyfish have evolved to be successful in their special environment in competition with bilaterians metazoans. Since no fossil evidence for the formation of these sensory organs in ancestral jellyfish is known it remains unclear when during their evolution such organs evolved and how close jellyfish or comb jellies with sophisticated sensory organs are in that respect to their ancestors that were not exposed to fast moving vertebrate, cephalopod, or crustacean predators that only evolved millions of years later.

Molecular data have clearly shown that both cellular compartments harboring these molecularly distinct response modules, as well as basic development of the organs containing specialized photo-or mechanosensitive cell types depend on closely related and yet discrete transcription factors that show in vertebrates some overlap in expression and function. Remarkably, these transcription factors are nearly identical in vertebrates and arthropods and seem to be identical for both organs in coelenterates. Hair cells in the vertebrate ear depend on the bHLH TF *Atoh1* but also express other bHLH transcription factors such as *Neurod1*. The vertebrate retina likewise requires several bHLH transcription factors for the development of different cell types: the *Atoh1*-like bHLH TF *Atoh7* for ganglion cells (some can respond to light) whereas others drive photoreceptor differentiation, mostly in combination with other factors. This similarity in the use of bHLH TFs in distinct neurosensory organs can either suggest that chance events randomly selected those transcription factors or that their expression reflects a deep common molecular ancestry. Below is the evidence supporting the second idea and arguing against the first.

Consistent with the overlapping use of bHLH TFs in cellular differentiation in vertebrate eyes and ears, the transformation of epidermal cells to be able to respond to these bHLH transcription factors with photo- or mechanoreceptor differentiation are organized in distinctly different organs by molecularly closely associated transcription factors, *Pax2/8* for ears and *Pax6* for eyes in vertebrates (Fritzsch and Martin, 2022). Interestingly, the lack of *Pax6* eliminates eye formation, whereas only the combined loss of *Pax2/8* abolishes ear formation indicating that gene multiplication in *Pax2/8* provides some signal redundancy. Interestingly, birds have lost *Pax8* and compensate with *Pax2* for the absence, supporting the notion of functional redundancy of *Pax2/8*. Likewise, a third transcription multiplication generated *Pax2/5/8*. Replacing *Pax2* with *Pax5* in mice shows that *Pax5* is functionally equivalent to *Pax2*. It is noteworthy that *Pax5* is only expressed very late in development in the ear of mice, suggesting that its expression evolved to be different and is used for other organs, whereas the signaling ability of the *Pax2/5/8* protein remained mostly conserved. Obviously, the duplication event of these transcription factors has not yet resulted in enough protein differences, allowing the *Pax5* protein to substitute for *Pax2* or *Pax2* to substitute for *Pax8*.

This finding is consistent with the more general idea proposed by E Davidson (Peter and Davidson, 2015) that evolutionary changes are prominently driven by altering the expression of duplicated genes, and altering the protein sequence of duplicated genes is a later step after they are differentially expressed. Sequence change might follow (or the gene might become a non-coding pseudo gene) with a delay after the segregated expression of the duplicated gene and its embedding in different developmental modules was accomplished. In addition to its function in ear formation, loss of *Pax2* also leads to the loss of ganglion cells in the retina, indicating incomplete segregation of the fate of eyes and ears in mammals. Yet another, even earlier expressed transcription factor, *Eya1* (eyes absent) leads to loss of eyes, ears, and kidneys when mutated. The effect of *Eya1* mutation shows that in the very early stages of development of sensory and excretory organs, the same factor may regulate comparable early differentiation steps of several organs in the same embryo, setting up their future distinct development that will be driven by progressively different downstream genes, starting from already distinct response properties to the same factor in different parts of the epidermis or the mesoderm. This cooption of novel kernels of regulatory components into gene regulatory networks (GRN) is apparently widespread in bilaterian evolution rendering the use even of genes that are homologous based on sequence similarity to establish across phyla organ homology extremely difficult. I will provide a few examples further below that indicate either common ancestry or cooption.

For the evolutionary developmental changes, I am aiming to illustrate here it is important to note that jellyfish have only a single ancestral *Pax2/6* gene that is expressed in both eyes and statocysts of jellyfish, presumably driving their development in analogy to the function of the two distinct *Pax2* and *Pax6* genes of mammals. Likewise, jellyfish have only a limited set of bHLH transcription factors, indicating that the formation of distinct organs with distinct sensory cell types able to perceive either light or gravity is possible even if these major factors are apparently identical. As with mammalian eyes, ears, and kidneys that all depend on *Eya1/Six1* for their initial precursor formation, other factors that interact with the single Pax and the few bHLH factor of jellyfish are needed to help understand how similar genes can mediate the differentiation of distinct organs with distinct cell types. These examples highlight that a more complete analysis of the development of diploblasts using modern molecular genetic manipulations is needed to dissect the trajectories of cell and organ evolution beyond our current limited understanding and provide the molecular context within which these genes can achieve their effects. In particular, misexpression of specific transcription factors from a given animal in the homologous or paralogous gene position of another animal can highlight the degree of conservation of a given GRN kernel. More developmental expression data are needed before the evolution of these complex sensory organs in comb jellies and jellyfish can be put into the context of what such manipulations have revealed about the development of bilaterian eyes and ears found in different animals such as flies and mice.

What is already clear, however, is that the complex sensory organs, in particular those dedicated to mechanical stimulation induced by altered position relative to earth gravity, exist in free swimming animals that are structurally so different that they may have formed independently nevertheless using common transcription factors to guide certain steps of their development. Examples of such GRN similarities without indication of organ homology are the expression of nested sets of genes that regulate proximo-distal development in arthropod limbs and posterior-anterior development in vertebrate brains. In each case, the basic expression that I refer to as 'GRN kernel' is supplemented by other transcription factors generating organs that are not homologous beyond that kernel. Likewise, anatomically different statocysts of jellyfish and comb jellies may be found to use transcription factors that are as related to each other as those of jellyfish seem to be to vertebrates but only for a comparable developmental step.

In this context, it is important to understand that even the structurally very dissimilar mechano- and photoreceptors of flies require the mammalian *Atoh1* related bHLH transcription factor *atonal* for their differentiation, clearly supporting the notion that to a certain extent, morphology is uncoupled from genes that are essential to guide specific steps of their development. Moreover, replacing the fly *atonal* with the mammalian *Atoh1* and vice versa leads to the normal development of fly eyes and chordotonal organs as well as mammalian hair cells and neurons. As outlined above, for jellyfish eye and statocyst development, yet incompletely understood molecular differences in the entire developmental GRN cascade provide a dissimilar enough context that allows differentiation of distinct sensory cells and sensory organ types using for a specific step of the developmental program the same or a closely related transcription factor.

That similar transcription factors can act across phyla was first revealed by the famous *Pax6* misexpression experiments showing a transcription factor that was able to organize eyes in different phyla, activating existing GRNs to drive host species specific organ development. For example, *Pax6* of vertebrates can induce compound eye formation in flies (but not vice versa). Likewise, *Pax2/6* of jellyfish can also induce eyes in flies, clearly supporting the notion that a given step in eye development of these structurally different eyes is dependent on a *Pax6*-like transcription factor. This similarity in the action of the same transcription factor in species with morphologically distinct eyes that evolved independently in each of these lines indicates a blurred line of the traditional homology concept. While specific transcription factors that are necessary for eye development are molecularly so similar that they can be replaced by each other, they nevertheless guide steps toward a morphologically diverse adult organ that evolved independently in each of these lineages. To drive this point home fully, the larval eyes of flies do not require *Pax6* whereas adult lateral eyes do. On the other extreme, while *Pax6* is expressed in the photosensory tips of starfish arms, starfish do not develop eyes or morphologically distinct photosensory cells.

What these complicated details show is that it remains unclear at the moment why certain GRNs are so conserved that a given transcription factor can substitute across phyla while the same factor may not be used in comparable organ development in the same species or may be expressed without driving organ development in other species. A given transcription factor may be a component of the evolutionary subfunctionalization of a specific receptor in different species or may be an integrated part of neofunctionalization participating in comparable steps of different sensory organ development in the same species. For example, the presence of the fly ortholog of *Neurog1* (tap) and *Pou4f3* (*acj6*) regulate fly olfactory receptor development, whereas they regulate in mice mechanosensory neuron and mechanosensory cell development. This could either indicate that in the common ancestor of flies/mammals, those factors were involved in chemo- mechanosensor development, and this function was differentially retained in either lineage. Alternatively, either function reflects cooption without any implication of homology of the respective organs beyond that molecular aspect of the GRN. As in the above outlined case of the Pax/bHLH gene cascade, more of the other modules into which these genes are embedded need to be known to decide between those two possibilities.

Combined, these data suggest that homology, even of organs with similar functions, does not follow from a single gene that is functionally equivalent, as demonstrated in replacement studies such as those conducted for *Pax6*. The molecular basis of the entire developmental GRN cascade of gene activation needs to be investigated to establish how many genes are similar in a given sequence and function in homologous organs such as vertebrate eyes and are different in non-homologous organs such as compound fly eyes. The already reasonably well-established interactions of *Pax6* with several other factors to establish a variable GRN network that shows differences between eye development of distant species clearly outlines the limits of contemporary attempts to use incompletely understood molecular data to refute or support homology based on organ similarities and known evolutionary history reflected in their systematic positions. Clearly, nobody would use the clear dependence of both kidney and ear development on the sequential activation of such powerful transcription factors as *Eya1/Six1* and *Pax2* to indicate the homology of mouse ear and kidney.

To elucidate this problem further, I would like to present a superficially similar case to the above outlined organ specifying GRN using the apparently simpler case of a cell fate specifying transcription factor. The bHLH transcription factor *Atoh1* of mice can functionally replace the fly bHLH TF *atonal*, generating photo- and mechanoreceptors of the fly type by activating fly specific downstream factors. Unusual for this kind of experiment is that the fly *atonal* can equally function in mice making hair cells in the ear and neurons in the brain, suggesting a high level of conservation of the regulation of the downstream GRN cascade that requires *atonal/Atoh1* for its activation. Obviously, *atonal/Atoh1* can drive mechanosensory development in flies and mice, indicating some conservation of *atonal/Atoh1* function despite the morphologically distinct mechanosensors that seem not to share similar mechanosensory channels and those channels are distributed to different parts of the mechanosensory cells (kinocilium in flies, stereocilia in mammals). In addition, *atonal* is involved in regulating photoreceptor development in flies but not in mammals whereas mammalian *Atoh1* regulates specific aspects of neurogenesis in most neurons in the mouse brain. Other functions of *Atoh1* in mice such as regulating the differentiation of Paneth cells in the intestine or Merkel cells in the skin clearly show that within a species a given transcription factor is tied into certain developmental steps of crossly dissimilar and clearly non-homologous cells in distinct organs. Therefore, while a given transcription factor and its experimentally conserved function indicate that a given step is conserved across organ development within and between animals, the complement of other genes and their level of conservation needs to be established to avoid premature suggestions of puzzling homologies across phyla.

As a rule, not a single gene but the expression context of a given transcription factor will determine how the function of that gene is used to develop a perhaps non-homologous cell or organ. In particular, very ancient TFs may be conserved for basic steps in a given cellular development such as apex/base specification in an epithelial cell (shared among all *Atoh1* expressing cells), generation of microvilli and kinocilia on one end, and of secretory vesicles on the other end of a given cell (again shared across all these cells during certain developmental steps). In conclusion, the conservation of a given TF in clearly non-homologous cell types may happen as a reflection of the von Baer principle, namely that general steps in cell or organ development come before specific steps (von Baer, 1828). Thus, a neuron is first an epithelial cell before it differentiates as a neuron and *Atoh1* (and other equally conserved transcription factors) can support such a developmentally early step, hence their conservation across phyla. Later I will discuss that modification of developmental GRNs can happen not only as a terminal addition but also as an intercalation of factors prior to the conserved factor. At a macroscopic level, such features of similarity have long led to the recognition of comparative development as an hour class where both very early and very late stages of development diverge more from each other compared to an intermediary, conserved stage.

Going beyond the molecular similarity as defined by the sequence identity of an orthologous, gene sequence are data generated by related but clearly different genes. In the case of *Atoh1* outlined above, three sets of genes are very similar and bind to closely comparable DNA sequences that differ only in a few nucleotides (Fig. 6) forming a triad of sister genes and sister binding sites. Such sequence similarity and established molecular ancestry that suggests duplication and diversification allow using those fairly similar genes, now used for distinctly different steps in the development of cells and organs, to test if they have enough overlapping signal capacity that they can substitute for each other in different cells of the same organism. Such data in the retina showed surprising effects whereby one gene may be able to substitute for another but not vice versa, indicating that the evolutionary alteration of different developmental trajectories is not happening at the same pace in different cell types. Since mutation rates are constant, this implies that evolvability differs between different cell types in the same organ in terms of GRN modifications.

We have begun such analysis generating mice in which *Atoh1* is replaced by the close, but functionally distinct mouse bHLH gene, *Neurog1*. Given the alleged inhibitory feedback role of *Neurog1* on *Atoh1* and vice versa, it could have been possible that simultaneous co-expression of both bHLH TFs from the two *Atoh1* loci block transcription of *Atoh1* expression so much that even mice heterozygotic for this construct cannot differentiate *Atoh1* dependent cells and thus die like *Atoh1* null mice. Indeed, sophisticated in vitro analysis shows that *Neurog1* can accelerate the inactivation of *Atoh1* that, combined with haploinsufficiency, could result in a lack of cellular differentiation in all *Atoh1* expressing cells and thus in the death of the animal. That did not happen! However, mice homozygotic for the replacement of *Atoh1* by *Neurog1* died at birth and showed almost no differentiation of cells normally dependent on *Atoh1*, resembling an *Atoh1* knockout mouse.

Closer examination using a more complicated construct in which *Atoh1* is only transiently expressed from one allele with *Neurog1* being expressed continuously from the second allele shows the limits of the needed similarity of closely related genes to substitute for and drive the differentiation processes normally mediated by *Atoh1*. Despite the expression of *Neurog1* in developing hair cells, *Neurog1* can only mediate the differentiation of dysfunctional hair cells that need *Atoh1* for an unclear initial step in the differentiation that cannot be substituted by *Neurog1*. This experiment also validated the conclusion derived from the mouse *Atoh1* replacement by the fly *atonal* gene that those two genes are indeed homologous. Obviously, they can catalyze specific steps in fly and mouse development and the sequence similarity reflects needed binding properties to mediate that specific function and are not exerting their specific function through activation of generalized bHLH binding sites. In contrast to hair cell precursors, neuronal precursors can easily be converted to hair cells by eliminating the transcription factor *Neurod1* which normally serves to suppress *Atoh1* expression in neurons. In *Neurod1* knockout mice many neurons differentiate as hair cells instead. Clearly, even in adjacent cell types derived from the common neurosensory precursor population of the ear, there is a local variation in the timing of expression of closely related bHLH TFs within the different steps toward neurosensory cell commitment and differentiation. *Neurog1* is establishing commitment and thus its absence can result in fate switching whereas *Atoh1* initiates differentiation after the commitment of the precursor has been established by other means.

In summary, detailed molecular analysis of transcription factors needed for sensory cell and organ development shows a certain level of conservation of genes that are interpreted as being derived from the same ancestral genes based on their sequence similarity (protein homology as defined by amino acid sequence similarity). Swapping those genes between distantly related animals sometimes shows functional conservation, apparently as far as conserved steps in sensory organs (*Pax6*) or sensory cell development (*Atoh1*) can be driven by the conserved part of such transcription factors. However, the outcome of such developmental replacement leads to host specific organs or cells due to the species-specific molecular context. Going beyond the initial excitement that such conserved genes could be used to define homologies across phyla, it now appears that more data on the molecular context within these homologous transcription factors signal is needed to establish or refute the homology of cells or organs. Simple speaking, a Paneth cell of the intestine, a hair cell of the ear, or a granule cell of the cerebellum, are not homologous simply because they express the same *Atoh1* gene at one step during their development, even if all of these cells share also some degree of mechanosensation. Clearly more molecular work at the level of temporal cell specific transcription profiling is needed before we can understand how a given developmental kernel or core regulatory complex of the cell developmental GRN ties into specific adult functions. It is conceivable that some kernels of a given GRN may differ between various cell types that may have coopted such a developmental step due to necessities that are now unclear. Establishing the homology of cell types across phyla based on molecular similarity is, in many cases, not yet possible due to limited data sets and an even more limited understanding of their functional significance.

Defining Novelty in the Neurosensory System

1.6. Chance and necessity in the formation of sensory neurons

Above I outlined how gene duplication and diversification of regulatory processes driving expression of duplicated genes in distinct cell types organized into discrete organs can lead to the formation of segregated sensory modalities using the example of mechanosensation and photic sensation. The formation of these 'sister' sensory cells and organs was enabled by duplicating 'sister' genes needed for organ and cellular development, all presumably tied into regulating the expression of the ancestral proteins that enable the perception of a specific sensory input, light, or mechanical stimulation. The selective advantage of segregating these light and gravity mediated stimuli to allow targeted movement within the water column to either move closer to the surface or deeper into the water for feeding on different prey may have provided an improvement for organizing those sensory inputs into distinct sensors for better-coordinated motor output. Both the **arrival of the fittest** (duplication of factors that allow segregation of sensory stimuli to distinct sensory cells in specific sensory organs) and the **survival of the fittest** (providing an advantage to orient during swimming in the water column using the antagonistic light and gravity inputs for orientation) are plausible for sensory organ evolution of jellyfish in terms of generating the cellular diversity and selecting the outcome. This basic selective advantage is, however, unclear for the duplication event that split a sensory cell with its own axon into two sister cells: a sensory cell without an axon and a sensory neuron conducting the information gathered by the sensory cell to the central nervous system such as the taste buds, inner ear, lateral line and electroreceptive sensory cells and their associated neurons of vertebrates and enigmatic sensory cells in tunicates. Below, I will highlight the molecular and neuroanatomical changes followed by speculation of selective advantage.

The idea (Fig. 7) that gene duplication and diversification enable the split of a single neurosensory cell that receives sensory stimuli with specializations of the apex and connections to downstream neurons with synapses formed at the axonal terminals into two cells, was initially proposed before the molecular basis of a sensory cell and the sensory neuron differentiation was established through mutagenic analysis of relevant transcription factors. Of the two logical possibilities of either the sensory cell or the sensory neuron being molecularly closer to the single neurosensory cell, it was initially proposed that the sensory neuron should be more similar. Subsequent analysis revealed that sensory neurons of the ear follow a molecularly distinct pathway for development that uses transcription factors that are not expressed in neurosensory cells. In

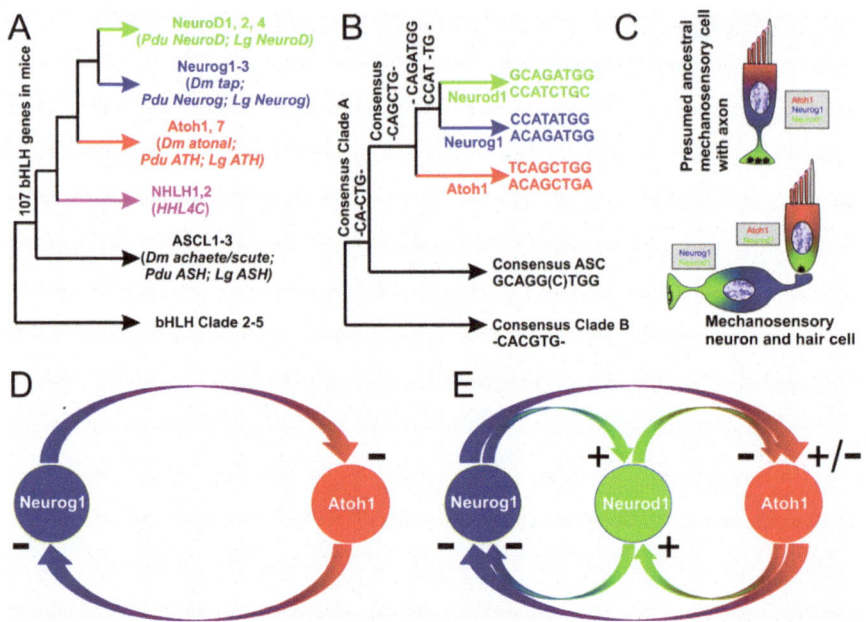

Figure 7: Molecular similarity of bHLH transcription factors and their E-boxes the bind to indicates likely multiplication and diversification at molecular and cellular levels. (A) Relationship between the bHLH transcription factors. (B) Relationship between the E-box binding sites of the various bHLH transcription factors. Note co-evolutionary changes in the bHLH transcription factors (A) and the E-box sequences (B). (C) The presumed ancestral mechanosensory cell with an axon may have contained three bHLH transcription factors: *Atoh1*, *Neurod1*, and *Neurog1*. Incomplete segregation of these three transcription factors results in a derived condition with separate hair cells expressing *Atoh1* and *Neurod1* and sensory neurons expressing *Neurod1* and *Neurog1*. Some experiments suggest that *Neurog1* and *Atoh1* are in a negative feedback loop (D). Knockin of *Neurog1* into *Atoh1* and effects of *Neurod1* deletion suggest that this feedback loop is more complicated and at least in part mediated by *Neurod1* (E). Taken from (Fritzsch and Elliott, 2017).

contrast, sensory cells use transcription factors similar to those used in neurosensory cells and such factors (*Atoh1, atonal*) can be replaced between fly neurosensory cells and mouse sensory cells. In contrast, the bHLH TF that is essential for sensory neuron development (*Neurog1*) cannot on its own substitute for the bHLH gene relevant for sensory cell differentiation in the ear (*Atoh1*) and is not used in fly mechanosensory development. This suggests that the mammalian sensory hair cell should be regarded as the cellular homologue of the fly neurosensory cell whereas the mammalian sensory neuron is the cellular equivalent of a highly derived cell that uses a distinctly different, presumably duplicated molecular

GRN sequence for its development. Despite anatomical differences in the mechanotransduction apparatus down to the mechanotransduction channels and their distribution in different cellular processes (stereocilia in vertebrates, kinocilia in flies), the fly *atonal* gene can direct mouse hair cell development whereas the mouse *Atoh1* can direct fly chordotonal cell development. Thus, while the molecular similarity of *atonal/Atoh1* allows a mutual replacement for the critical developmental step in either cell type, neither topology nor functional details are conserved. These turns fly chordotonal organs and vertebrate cochlear hair cells into serial sister homologs that have retained only specific core regulatory components of their developmental GRN, intercalating upstream regulators driving their topology and substituting downstream regulators for morphological/molecular functions presenting a cellular equivalent of the hour class of overall organismal evolution and development.

The complexity of serial sister homology to establish evolutionary and developmental similarities is particularly obvious in cases of shared labor being subdivided into two evolutionary and developmentally related, but morphologically and functionally distinct cell types. Such a process recapitulates the basic initial diversification of cell types upon formation of the first metazoan forming the first two distinct cell types instead of morphological and functional identical clusters of cells in a colony. The formation of second order sensory cells and their sensory neurons can allow analyzing how gene multiplication of crucial transcription factors and tying them into different downstream modules that were associated with the ancestral neurosensory cells may have been accomplished, shedding light on steps taken in ancestral metazoan cell diversification now ~750 million years in the past.

Apparently, simple duplication of a single crucial transcription factor is too unsophisticated an approach as it does not allow fine tuning the reciprocal suppression of these transcription factors to drive distinct differentiation of either of the two sister homologs. I will limit the consideration here to only three closely related bHLH TFs, *Neurog1*, *Neurod1,* and *Atoh1* (Figs. 5, 7). While some aspects of the complexity of their interaction have been revealed experimentally, I posit that the expression of other bHLH TFs in this system is underrated as their function is camouflaged mostly by the prominent transcription factors currently partially understood. Before I explain the necessary steps that need to be taken to evolve two sister cell types out of one ancestral cell type I will present what experimental analysis of various gene functions in this system has revealed.

Work on mutant mice has established that the bHLH TF *Atoh1* is essential for hair cell differentiation and parallel work showed that the bHLH TF *Neurog1* is essential for sensory neuron development.

Importantly, the homolog of vertebrate *Neurog1* in flies (tap) is not expressed in chordotonal organs, indicating that flies go a different route with newly evolved bHLH TFs to drive cellular differentiation of the specific fly type. Moreover, a third bHLH TF (*Neurod1*) is also expressed in sensory neurons and most die in mouse mutants lacking that gene whereas flies have apparently lost the *Neurod1* gene that is, however, present in basal bilaterians. *Neurod1* is also expressed in hair cells and changes the phenotype of some hair cells if absent. For sensory neurons, their fate hinges on the presence of other bHLH TFs that can either rescue neuronal differentiation or convert already delaminated sensory neurons into hair cells. The conversion of sensory neurons into hair cells is mediated by the transient expression of *Atoh1* that is normally suppressed by *Neurod1*. In the absence of *Neurod1*, *Atoh1* remains upregulated downstream of *Neurog1* and converts some neuronal precursors into hair cells. Substantial in vivo and in vitro evidence shows that *Neurog1* and *Atoh1* are in an inhibitory feedback loop but not all effects of loss of *Neurog1* are directly related to the disinhibition of *Atoh1*. For example, loss of *Neurog1* leads to premature hair cell differentiation through upregulation of *Atoh1* (utricle and apex of the cochlea), leads to the formation of hair cells in areas of the ear where normally none develop (ductus reuniens, greater epithelial ridge) but can also truncate the formation of hair cells (saccule, organ of Corti). These regional specific effects on the fate of possibly lineage related cells cannot be explained by a simple negative feedback loop between *Neurog1* and *Atoh1*. At least one additional factor (more likely several) is needed to achieve the apparent differential effect of *Neurog1* deletion on precursor cells of the ear. In other words, temporal and spatially distinct contexts of transcription factors will drive local variations of common gene modification effects.

It is difficult to imagine a scenario whereby a simple gene duplication and the establishment of a negative feedback loop to ensure differentiation into either one or the other type of cell can evolve as it would generate an oscillating system of differential cell specification. By necessity, the initial duplication would have two genes coding for a similar protein but being driven in their expression by a subset or regulatory elements. Given the signaling redundancy of the two proteins, such duplication can only lead to selective advantage if the co-expression in the same cell is changed in intensity to adjust for gene doubling or altered timing deflecting overlap. Sequential activation, particularly if combined with negative feedback loops, can result in differential expression in daughter cells during asymmetric divisions, giving rise to a differentiating cell and another cell that repeats cell division. Extensive data support the notion that waves of gene activation occur during development, each leading to a modification of gene expression profiles, stepwise altering the expression profiles of precursor cells. While we tend to think of cell fate changes in terms of switches, in reality, it functions more like two dimmers where one is downregulated progressively to eliminate that potential fate while the other is upregulated progressively to establish a different fate. Simple gene duplication, while superficially appealing, lacks the subtlety needed to accomplish this incremental transformation of precursors.

Consistent with this scenario is that *Neurog1* can alter the level and thus signal strength of *Atoh1* in vitro, possibly through accelerated degradation of the mRNA. In contrast to these in vitro data, in vivo data of co-expression of *Neurog1* under *Atoh1* regulation shows that there is hardly any effect on the *Atoh1* level and thus function and normal hair cell differentiation. In addition, certain effects on the truncation of hair cell formation are similar in *Neurog1* and *Neurod1* mutants, suggesting that some of the *Neurog1* effects are mediated by *Neurod1*. Clearly, the lack of *Neurod1* causes premature upregulation of *Atoh1* and disinhibits hair cell fate suppression in sensory neurons. How much of the apparent action of *Neurog1* is due to the upregulation of *Neurod1*, and *Neurod1* mediated suppression of *Atoh1* relative to *Neurog1* mediated suppression, requires further investigations. In essence, at least for vertebrate mechanosensors and sensory neurons, it appears that the ancestral neurosensory cell evolved several, sequentially activated bHLH TFs that each evolved to regulate a different module of neurosensory cell development. Once the within-cell compartmentalization had evolved, the modular differentiation could be parceled to differentiate distinct cell types through the exclusion of specific modules driven by genes not expressed during that specific differentiation. Re-expressing those genes (as in the case of *Neurod1*) leads to the adoption of a phenotype related to those specific modules that would be expressed. Logically, this should always be later differentiating phenotypes as is the case in the mammalian ear where early forming sensory neurons turn into late forming hair cells. Such transformation should not be possible if an early expressed bHLH factor replaces a late expressed bHLH factor. Indeed, expressing *Neurog1* instead of *Atoh1* dose under no circumstance changes the phenotype of the late developing hair cell into an earlier developing neuron. A most extreme, not yet tested temporal misexpression would involve using a very late bHLH TF such as *Neurod1* and expressing it ahead of all other bHLH TFs. Such early misexpression will likely result in catastrophic derailment of ear development as the cells simply are not ready to cope with a late expressed transcription factor. What is clear in mice is that mutations of *Neurod1* indicate the asymmetric dependency of sensory neurons/hair cells on this transcription factor in addition to its ability to negatively regulate both *Neurog1* (for sensory neurons) and *Atoh1* (for sensory hair cells): Most sensory neurons are lost or converted to hair cells in *Neurod1* null mice. In contrast, only a limited number of hair cells do not develop but several outer hair cells in the apex change their phenotype due to premature upregulation of *Atoh1*. This asymmetry is important as it advances the simple negative feedback loop possible with a *Neurog1/Atoh1* feedback loop and avoids an oscillatory behavior inherent to such loops.

Irrespective of our current lack of detailed understanding of the origin of the sister cell types of hair cells/sensory neurons in the ear, all vertebrates have mechanosensory cells connected to the brain via sensory neurons. However, the evolution of sensory neurons and sensory cells from the ancestral neurosensory cell may apparently already have happened in tunicates. Molecular evidence supports the notion of homology of sensory cells, but it remains unclear if sensory neurons of tunicates are homologs of craniate vertebrates due to incomplete data on their molecular signature. Finding that the molecular signature of neurosensory development is identical in tunicates and vertebrates as much as the sensory cell development had similarities, would strengthen the case that the formation of sensory cells and sensory neurons out of neurosensory cells predates vertebrates. What remains unclear is the selective advantage of this arrangement in part related to an incomplete understanding of how many sensory neurons innervate how many sensory cells in tunicates. Assuming that sensory neuron evolution is ancestral to tunicates and craniates and is clearly absent in the outgroups of chordates and hemichordates, more data on tunicate sensory neuron convergence/divergence to sensory cells is needed to understand the selection advantage offered by this arrangement.

In vertebrates, the number of hair cells in most sensory epithelia is higher than the number of nerve fibers innervating them, and this tends to change over time due to more proliferation of hair cells relative to neurons in most anamniote vertebrates. As outlined above, the ratio of sensory neurons to sensory cells is unknown in tunicates. Knowing this ratio could show that convergence of several sensory cells onto a single sensory neuron could increase sensitivity. Even though, it remains unclear what the advantage of such an arrangement would be relative to converging multiple neurosensory cells on the same information processing second order neuron in the CNS. While ultimately the formation of a sensory cell/sensory neuron configuration instead of a neurosensory cell increased the 'evolvability' of various species and end-organ specific convergence/divergence ratios, it remains unclear what the immediate selective advantage may have been for the chordate ancestor that evolved that configuration. Duplication of whole cell fate developmental GRNs followed by a mutation leading to the formation of a novel cell type must have happened. Such molecular duplication leading to multiplication of bHLH transcription factors also multiplied downstream genes such as the *Pou4f* factors. Vertebrate hair cells activate a sequence of *Atoh1>(Neurod1)>Pou4f3* whereas sensory neurons activate the sequence of *Neurog1>Neurod1>Pou4f1*. Even further downstream are distinct activations of the neurotrophin *BDNF/Ntf3* in hair cells and the neurotrophin receptor *TrkB/TrkC* in sensory neurons. It is conceivable that the whole genome duplication known for ancestral chordates has generated the duplication events that led to the selection of different paralogous genes to support the differentiation of two distinct cell types with a life supporting interaction between them. It appears that *Pou4f1* is the paralogue that evolved a novel function in chordates whereas *Pou4f3* is the orthologue that is also present in flies. It seems logical that cell duplication combined with shared labor through the allocation of entire cell differentiation modules to sister cells can only happen after the molecular pathway within a single cell was duplicated, tying sub functionalized transcription factors into directing distinct modules. Once that is achieved, specific aspects of such a developmental sequence can be relegated to different cell types that will eventually become more distinct as evolution progresses. While the arrival of distinct molecular pathways through gene duplication is consistent with molecular data of chordates, explaining the arrival of the fittest, the immediate selective advantage of the sensory neuron split into distinct cell types (survival of the fittest) remains obscure.

The evolution of the sister cells in ancestral chordates could possibly be resolved through a better understanding of the development and functional significance of both primary neurosensory with their axon as well as a combination of sensory cells and their information conducting sensory neurons in cephalopod mollusks. This mix of neurosensory cells, sensory cells, and sensory neurons might have been present in ancestral chordates before the segregation of sister cells was completed.

1.7. Lessons learned: Transforming existing molecular systems of single celled organisms to open new windows to sense the world in multicellular organisms using discrete sensory channels

This chapter started with the premise that molecular evolution predates cellular evolution. The evidence provided highlights that transmembrane channel protein(s) that could be transformed through mutation into the vertebrate mechanosensory channel exist in single celled organisms. Likewise, cadherins that enable cell-cell interactions to allow metazoans to evolve into tip links for mechanosensors predate metazoans and various vesicular release mechanisms and ion selective channels. Single cells have a multisensory capacity, responding to certain mechanical, photic, and chemical stimuli by affecting movement direction and intensity. Evolving the generalized sensing into a more restricted cellular sensing allocated to distinct receptor cells specialized for mechanosensation such as sound, the photic sensation of specific wavelength, or chemosensation involving specific smell or taste requires primarily segregation of those receptor molecules into discrete sensory cell types while at the same time suppressing sensing of other modalities. Once the cell type specific distribution of molecules allows sensing a specific modality, this sensing can be improved by adding stimuli enhancing features such as is possible in various sensory organs harboring these cell types. Parcellation of general sensory mechanisms driven by discrete molecules within a single cell into segregation of the regulatory machinery to allow predominant followed by exclusive expression of a given sensory molecule in a given sensory cell was possible after multicellularity evolved, essentially setting up a runaway selection that drives evolution of specialized receptor cells as sister cell types of the original pluripotent single cell ancestor.

Transcription factors that regulate the transition from vegetative to generative phases in single cells, such as bHLH factors, were apparently recruited to tie into the differential regulation of sensory molecules. Once the multiplication of such factors allowed for regulating sensory molecular development, it could become the basis of sensory modality segregation. This process is particularly obvious in the sensory cell connection of photic and mechanosensors where multiple transcription factors form diverging cascades to segregate distinct sensory cell types into discrete organs. The degree of segregation of these processes can be mostly overlapping, as in flies or is nearly completely segregated, as in vertebrates.

A distinct feature of this later sister cell formation is the formation of sensory cells with no axon and matching sensory neurons conducting sensory cell information to the central nervous system in chordates out of a neurosensory cell with its own axon in chordate ancestors. While superficial similarity to the sister sensory cell types that house distinct sensory modalities, the segregation, in this case, generates two types of cells with very different specializations: sensory perception and sensory information conduction. Despite this obvious difference, the basic principle of transcription factor multiplication followed by specification to regulate only distinct cell fate processes are, in principle, identical. Thus, the molecular diversification and tie into the sensory and neuronal feature of the ancestral neurosensory cell must be segregated inside this cell before two distinct cell types can be generated from a single precursor.

Notably, the hair cell afferent synapse is molecularly distinct from other chordate synapses in that it forms a presynaptic bar of various dimensions and uses otoferlin instead of *Munc* for synaptic vesicle docking. Similar synaptic bars are only found in vertebrates in photic sensors (rods and cones) indicating that this transmitter release mechanism may be ancestral to both single celled mechano- and photoreceptors. Interestingly, this cell type segregation seems to occur as a temporal progression using negative feedback loops of molecular orthologues to suppress alternative cell fates. Because of this progressive molecular segregation of cell fate, misexpression by various means of late expressing transcription factors can change the fate of sensory neurons into hair cells. In contrast, the expression of neuronal specifying transcription factors in hair cell precursors cannot transform them into neurons but cause aberrant hair cell development. What remains unclear is what the initial selective advantage of a sensory neuron/sensory cell pair could offer over a single neurosensory cell. This is particularly problematic given that any synapse will cause a delay of information flow that requires very fast neuronal conductance and thus myelin around sensory neuron axons that evolved among chordates only in vertebrates whereas the neurosensory sister cell pair formed in ancestral vertebrates and possibly already in the common ancestor with tunicates.

There is a rapidly expanding insight into cell fate specification and the generation of two distinct cell types for sensory and conductive function. However, there is also the caveat of a lack of understanding of the selective advantage of the initial formation of the neuronal/sensory cell instead of a single neurosensory cell. With both aspects in mind, I will next investigate how sensory cells became aggregated into multicellular sensory organs. It is possible that the segregation of the two sister cell types is tied into sensory organ formation as in cephalopods where the statocyst consists of a mix of neurosensory cells as well as sensory cells connected via sensory neurons to the central nervous system. If so, discrete sensory organ formation may have only offered a selective advantage through the formation of sister cell pairs of sensory cells/sensory neurons. The convergent formation of such pairs in cephalopods and vertebrates hints at an as yet incompletely understood selective advantage.

In the next chapter, I analyze the origin of the sensory neurons, focusing on the hair cells that receive the sensory neurons and connect neurons to the brainstem. I will compare the/hair cells/that are unique to the ear and related sensory cells first, followed by the diversity of the formation of the auditory system. Logically, the start of hearing was in sarcopterygians from which I can follow the generation of the mammalian auditory neurosensory development that reaches out to the forebrain to provide human ability, like talking and singing. In the end, I will briefly provide an overview of cross-talks between the auditory cortex that interacts with a large area of other inputs. Further, I cite the most recent data that are relevant for the vestibular, lateral line, electroreception, and, in particular, the auditory system.

CHAPTER 2

Evolving Mechanosensation for Gravity and Angular Orientation, Lateral Line, and Electroreception

Logically, one would assume that a transition of neurosensory cells replaced a neuron and a sensory cell, and eventually connected to a newly developed brainstem. Cephalopods show the co-existence of neuron/'hair cells' next to neurosensory connections but must be considered a parallel evolution of cephalopods (Figs. 4, 6, 8). In contrast, all taste, vestibular, lateral line, and electroreception are found in vertebrates and have an addition of the Merkel cells as separate trigeminal information. Olfactory senses are distinct and have a neurosensory connection from primary sensory neurons, projecting to separate second order connections, the olfactory neurons reach out to the olfactory bulb. Given that we have clarification of cyclostomes and that the gnathostomes are closely related with the four connections provided by taste, vestibular, lateral line and electroreception with different details implies an old separation of these sensory cells out of the spinal cord-like organization. In contrast to original ideas, it was viewed as a clade, starting with hagfish followed by lampreys and gnathostomes. In fact, recent evidence suggests that they are now viewed as a derived organization of hagfish that distantly relates to lampreys as a group of cyclostomes. Lampreys make a close relationship with a gnathostomes concerning the eyes, ears, lateral line, and electroreception. Distinct from taste buds (different genes of *Neurog1* versus *Neurog2*) defines all neuronal sensory neurons that are ending close to the central projections, distinct also from central projections of taste buds (solitary track compared to all mechanosensory and electroreception projections) and continued with trigeminal projection interacting with the spinal cord.

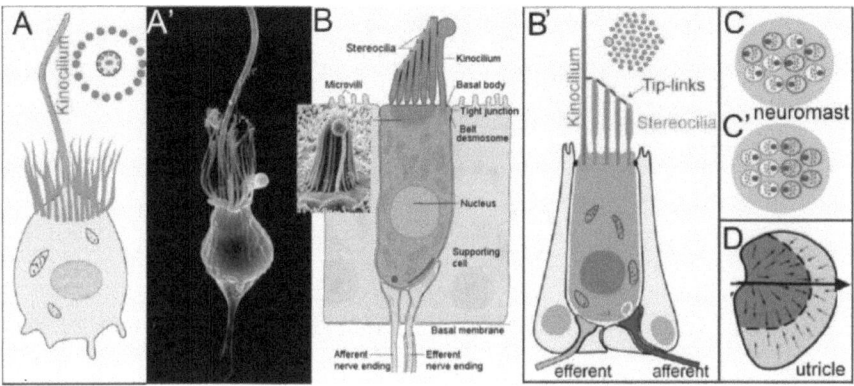

Figure 8: Cell morphology in single-celled choanoflagellates (A, A') and the vertebrate inner ear compared. SEM image (B) and the lateral line (B') with a choanoflagellate and the central kinocilium, the collar of microvilli surrounding the kinocilium, and a process to the opposite side of the cell to attach to the substrate (A, A'). The vertebrate vestibular hair cell has graded stereocilia asymmetrically organized relative to the kinocilium (B) comparable to the lateral line (B'). Stereocilia are connected by tip links to open ion channels. Instead of a process emerging from the opposite pole, hair cells receive afferent and efferent fiber innervations. The insert shows a frog hair cell with a kinocilium on one side of the stereocilia. Note that the different hair cells are forming into two polarities in the lateral line (C, C') and the utricle (D). Image combined from (Engelmann and Fritzsch, 2022; Fritzsch, 2021; Fritzsch and Straka, 2014).

An unclear problem of the four mechano- and electroreception (vestibular, lateral line, electroreception, auditory) is shared among all vertebrates: all representatives of all vertebrates have these neurons, have unique sensory cells, and have distinct central projections that have been implied to originate at about the same time in vertebrates some 450 Myo ago. In addition, a triple dependency is demonstrated in neurons, brainstem nuclei and sensory cells: neurons depend on sensory cells and vice versa, in addition, central nuclei depend on neurons and vice versa.

1. Neurons require *Eya1/Six1, Sox2,* and *Neurog1* to make all sensory neurons and guide their final differentiation under *Neurod1, Nhlh1 + 2, Pou4f1* and others, including *SWI/SNF*, to drive the central projection (*Pou4f1* and others) to their respective brainstem nuclei and guide the innervation of the peripheral hair cells. Lack of neurons reduces hair cells and results in the loss of many brainstem nuclei, depending on both ends of neurons of hair cells and central nuclei.
2. Several brainstem nuclei depend on *Lmx1a/b, Gdf7, Sox2, Atoh1, Neurog1/2, Ascl1, Neurod1, Phox2a/b, Ptf1a,* and other transcription factors are needed alone, or in combination, to form vestibular nuclei

(VN), intermediate nuclei (IN), dorsal nuclei (DN) and, in derived tetrapod's, cochlear nuclei (CN), add and replace the intermediate/ dorsal nuclei by the 'auditory nuclei'. A selective loss of nuclei without a loss of neurons and hair cells has not been investigated except for the unclear role of *Lmx1a/b*, and downstream genes of *Gdf7, Bmp's,* and *Wnt's* in auditory and vestibular nuclei. Loss of various nuclei is in part dependent on various nuclei. Even without *Atoh1*, which eliminates cochlear nuclei and cochlear hair cells, has normal spiral ganglion neurons that connect transiently the absent hair cells with the equally absent cochlear nuclei, indicating these independencies of neurons and suggest a distinct role of a later expansion to add central nuclei and peripheral hair cells. Loss of central nuclei will follow the absence of neurons but will be to a variable degree of ablation of all hindbrain nuclei.
3. Hair cells depend on *Eya1/Six1, Sox2,* and *Atoh1* in collaboration with *Neurog1, Neurod1, Pou4f3/Brn3c, BMP4, Shh,* and others to drive the differentiation of three types of cells: mechanosensory hair cells of the vestibular and lateral line, electroreception of hair cells, and the auditory type of tetrapod hair cells. Loss of hair cells results in loss of neurons and, consequently, a loss in brainstem nuclei.

The triple mutual dependency of neurons, hindbrain nuclei, and hair cells depends on selective connections of neurons to project hindbrain nuclei and hair cells. Experimental evidence is a central dominance of neurons and requires the brainstem and hair cells, eventually. Following the common origin of *Neurog1/Neurog2* for all neurons must be connected first with the various brainstem input to segregate trigeminal, vestibular, taste, lateral line, electroreception, and auditory nuclei. In contrast, the trigeminal innervation was originally with limited specialized Merkel cells but was supplemented by taste sensory cells, driven by *Neurog2*, and the vestibular/lateral line/electroreception/auditory hair cells. Given the sequence of nuclei > brainstem > hair cells are the sequence of development that is obvious in numerous proliferation studies and sequential proliferation.

The greatest problem is the following: How, in fact, do we know about the sequence of gene evolution of the four sets of neuronal, nuclei and sensory innervation develops? What do we know about the sequence of four distinct evolutionary primary projections (vestibular, lateral line, electroreception, auditory) and compare that to their sequential or simultaneous appearance of cyclostomes and gnathostomes (Fig. 9)? Arguably, the apparent simulation of three projections (vestibular, lateral line, electroreception) was well-known before the 'octavolateralis' was coined, used, and goes back before selective projection studies. Without proper tracing of central projections, many people adopted the 'octavolateralis' idea and assumed that all fibers are intermixed for

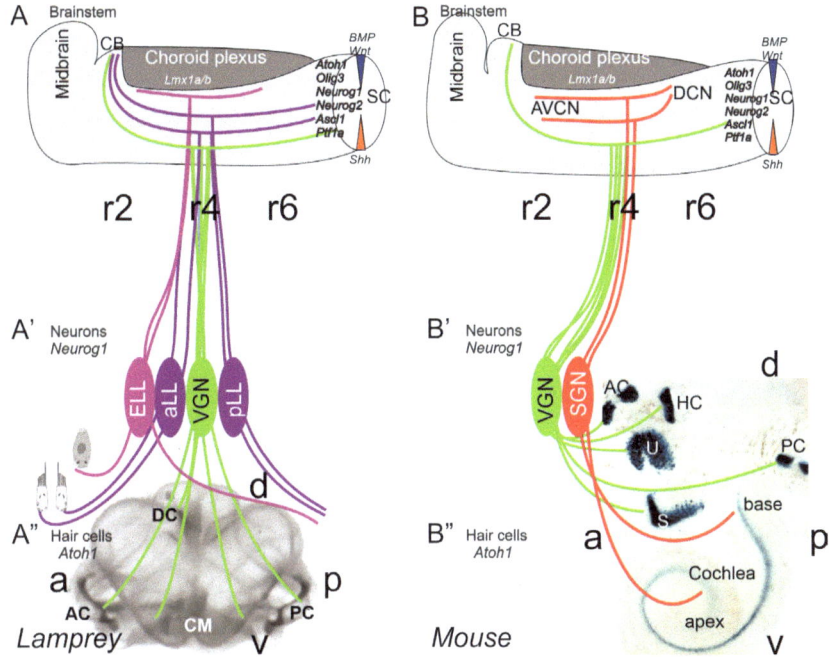

Figure 9: The brainstem is innervated by electroreceptor (ELL) and lateral line fibers (LL) that extend to innervate migration populations of LL and some ELL. The ear is unique in vertebrates, which gives rise to the VIII ganglia that innervate more ventral nuclei compared to LL and ELL projections to reach *Atoh1* (A). Neurons (*Neurog1*; A') form vestibular ganglia (VGN) to reach out 4 hair cell organs in lampreys (A"). A separate lateral line (LL) and electroreceptor neurons (ELL) that innervate hair cells project more dorsal in lampreys. Central projection depends on *Atoh1* to receive LL and ELL fibers, whereas several bHLH genes (*Neurog1/2, Olig3, Ascl1, Ptf1a*) receive all VGN (A). In the absence of ELL and LL development in amniotes, mammals develop separate spiral ganglion neurons (SGN; B') that extend from the cochlea (B") and end in a topological central projection that depends on *Atoh1* (B). The formation of VGNs (*Neurog1*; B') reach the 5 hair cells (B") to extend the distribution of bHLH genes. Note that certain areas are lost or gained which enter central projections near r4. Images are shown by miR-183 ISH (A") and *Atoh1-LacZ* (B"). AC, anterior crista; AVCN, anteroventral cochlear neurons; CB, cerebellum; aLL, pLL, anterior/posterior lateral line neurons; CP, choroid plexus; CM, common macula; DC, dorsal crista; DCN, dorsal cochlear neurons; HC, horizontal crista; PC, posterior crista; r2/4/6, rhombomeres; S, saccule; SC, spinal cord; U, utricle. Taken from (Fritzsch, 2021).

the central nuclei, something that is in stark contrast to the selective projections that show distinct projections of different sensory projections to three layers of mechano- and electroreception to their specific central targets, with the loss of lateral line and electroreception and gain of an auditory projection. Finally, there is a distinction between the spinal cord that differs from the brainstem: instead of a roof plate the brainstem is characterized by the choroid plexus.

In fact, recent tracing of individual projections shows that evolution forms sequentially and the addition of central projections connecting, while 'hair cells' was unclear among ancestral vertebrates. More recently, several data have now clarified the sequence of distinct projections and demonstrated a sequential development of fibers following a general central projection of neurons to their targets: the sequence of trigeminal > vestibular > lateral line > electroreception projects and has no overlap of individual nuclei projections, despite being all dependent on the same neurons driven by *Neurog1*. Fuzzy definitions of 'octavolateralis' and 'cochleovestibular' fibers fall apart once the distinct projections can be traced from distinct neurons and should eliminated this fuzzy definition by replacing 'octavolateralis' with vestibular, lateral line and electroreception and replace 'cochleovestibular' by vestibular and cochlear distinct projections in amniotes.

What we know is that all craniate vertebrates have to develop an otocyst that forms from a dorsolateral placode to generate mechanosensory hair cells and receives from a unique set of neurons derived from the inner ear that projects to a distinct central projection to the vestibular nuclei. A comparable set of nearly identical lateral line mechanosensory hair cells connected via distinct sensory neurons (anterior and posterior lateral line neurons) to project to reach distinct intermediate nuclei. Likewise, electrosensory cells are innervated by distinct fibers to receive a selective group of neurons that project to the dorsal nuclei. Finally, multiple and distinct evolutions of the various auditory sensory cells project the unique neurons to develop a distinct cochlear-like projection in all tetrapods, except certain amphibians who have reduced or lost the auditory input.

In contrast, chordate outgroups have only molecular features resembling placode formation and have distinct innervation by direct innervation from neurons, like the placodal derived vertebrate. Importantly, neither a selected formation of a distinct set of 'spiral ganglia' is absent in chordates of 3100 tunicates and 31 lancelets. Despite these molecular developmental similarities and the use of clearly homologous transcription factors for comparable steps in development, it remains unclear if the multicellular organs in the lancelet (organs of de Quatrefages) or the multicellular organs at the syphon of certain tunicates are parallel formations to the ear or lateral line of craniates or are precursor organs. The cellular similarities, including transcription factor expression in these putatively mechanosensory cells, are of no help as they could be recruited based on the ancestral history of the sensory cell evolution and not on the sensory organ history (Fig. 9). Moreover, since anatomically distinct and yet molecularly somewhat overlapping sensory organs are found in

Evolving Mechanosensation for Gravity and Angular Orientation...

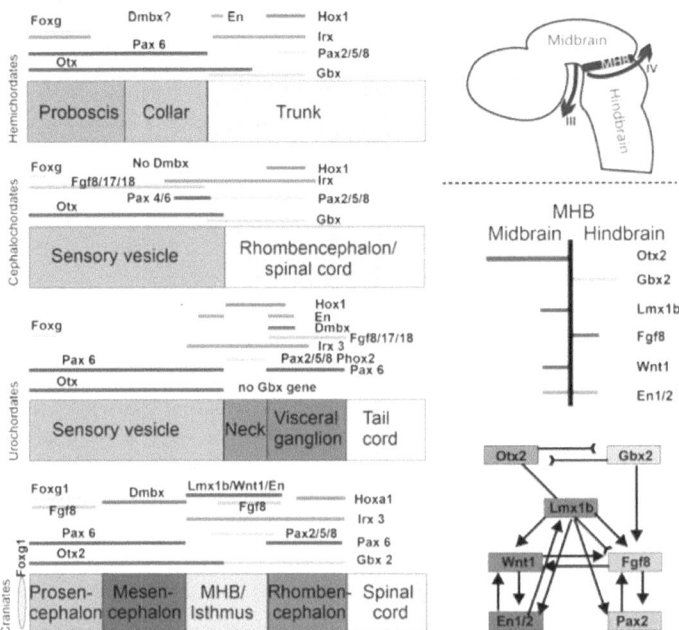

Figure 10: The evolution of gene expression at the midbrain–hindbrain boundary (MHB) is shown for deuterostomes (left). The MHB of vertebrates exhibits abutting domains of *Otx2* and *Gbx2* expression. This stabilizes the expression of Fgf8 (far right), which in turn stabilizes the expression of Wnt1 and engrailed (*En1*). Mutation of *Otx2, Gbx2, Fgf8,* or *Wnt1* eliminates the MHB. *Pax2/5/8* are also expressed at the MHB, whereas the expression of Dmbx occurs immediately rostral to the MHB in the midbrain to later expand into the hindbrain and spinal cord (bottom left). Note the partial overlap of *Pax2/5/8* with the caudal expression of *Otx2* and the rostral expression of *Gbx2*. Hemichordates (top left) have overlapping expressions of *Gbx, Otx, Irx,* and *En* in the rostral trunk. *Pax6* abuts *Gbx2* whereas *Pax2/5/8* overlaps with the caudal expression of *Gbx2*. Outgroup data suggest that coelenterates have a *Dmbx* ortholog, thus raising the possibility that hemichordates also have a *Dmbx* gene. Cephalochordates (top second left) have no Dmbx expression in the 'brain.' The *Otx* expression domain abuts the *Gbx* expression domain, as in vertebrates. However, *Gbx* overlaps with *Pax2/5/8* and most of *Irx3*. Urochordates (bottom first left) have no *Gbx* gene but have a *Pax2/5/8* and *Pax6* configuration comparable to vertebrates. *Dmbx* overlaps with the caudal end of the *Irx3* expression whereas *Dmbx* expression is rostral to *Irx3* in vertebrates. Together, these data show that certain gene expression domains are topographically conserved (*Foxg1, Hox, Otx*), whereas others show varying degrees of overlap. It is conceivable that the evolution of nested expression domains of transcription factors is causally related to the evolution of specific neuronal features such as the evolution of oculomotor and trochlear motoneurons (right) around the MHB. Experimental work has demonstrated that the development of these motor centers depends on the formation of the MHB. Modified after (Glover et al., 2018).

coelenterates that use homologous developmental transcription factors, it seems most likely that the craniate mechanosensory organ formation (ear, lateral line and electroreception) is a unique craniate feature with incomplete similarities in tunicates and, even less so, with lancelets (Fig. 10).

Generally speaking, making an organ system requires the formation of a precursor population that can be specified to differentiate as a sensory organ while allowing the precursor population to multiply and expand existing projections. Obviously, the common mechanical conservation of lateral line and vestibular information provides an overview of these similar mechanoreceptor cells that are common mechanosensory hair cells, closely of central projections to the brainstem, and allows common second order projections among these two sensory systems. In contrast, derived features set aside from the electroreception (electroreceptors, distinct central projection to the dorsal nuclei short projections exclusive from second order projections) and the auditory mammalian system (distinct organization of unique hair cells that project with a unique group of sensory neurons that project a unique second order projections distinct from many but not all other projections). All are driven by the dorsal/ventral separation dependent on *Shh* (central) and *Bmp's* and *Wnt's* (dorsal) that defines the roof plate by *Lmx1a/b* and *Gdf7* genes and defines the floor plate by *Shh* (Fig. 10).

I will provide more evidence to support those suggestions after introducing the molecular transformation of the hindbrain (Chapter 4) to accommodate the discrete projections of sensory neurons from the inner ear, lateral line, and, when present, electroreceptive, and auditory organs. In every case, I start with the neuronal projection first, followed by central projections to nuclei next, and followed by the hair cell formation last. Chapter 5 deals with the higher central auditory system.

2.1. Vestibular sensory neuron development requires a unique projection to distinguish brainstem from trigeminal projections and is the first hair cells to develop

An overview of vestibular neurons will depend on the inner ear with incomplete segregation of different neurons that reach the partial overlap of vestibular nuclei but are characterized as selective hair cells to two distinct inputs, the two to three cristae for angular input and the single (hagfish, lampreys) and up to three gravity input in gnathostomes (utricle/saccule/lagena).

Compared to the vestibular neurons that show a limited change of innervation, the ear is shaped drastically between the single circle of hagfish, compared to the complex organization of lampreys (Fritzsch et al., 2023a) and shows mostly a limited change of sizes among gnathostomes, mainly associated with the gain and loss of the size differences and the absence or presence of the lagena. The various organization of the ear among vertebrates is in stark contrast to the presumably ancestral connections of vestibular nucleus (composed of several subnuclei) and the central organization of vestibular fibers (Glover and Fritzsch, 2022).

2.1.1. Vestibular neurons

All inner ear neurons form from vestibular sensory neurons and depend on *Neurog1*. Elimination of *Neurog1* affects the central projection by resulting in loss of hair cells but is thus far unstudied on the effect of vestibular nuclei (Sun et al., 2022). The entire dependency extents from a single gene of the brainstem (*Neurog1*) but is differently added by *Neurog2*, but *Neurog1/Neurog2* are sequentially genes in the spinal cord whereas the *Neurog1/Neurog2* is separated into distinct expressions of proximal *Neurog1* and distal *Neurog2* expression in the brainstem (Glover and Fritzsch, 2022). In the spinal cord in lamprey and hagfish are characterized by distinct and early Rohon-Beard and dorsal cells forming a unique and early neuronal projection prior to the later developing spinal and branchial neurons, similar to lancelet's unique formation of peripheral innervation. In contrast to these central dorsal neurons of hagfish, lampreys and gnathostomes, including aquatic tetrapods like frogs, all neurons provide first a central projection to the vestibular nuclei followed by the expansion of peripheral processes to hair cells that will navigate to distinct targets: the different gravity receptor(s) composed of a single partially segregated gravity receptor in hagfish and lampreys, two (utricle and saccule, gnathostomes) or three (utricle, saccule, lagena) and have two (anterior and posterior cristae, hagfish and lamprey) or three canals (anterior, posterior and horizontal cristae; gnathostomes) for angular innervations. Nuclei partially overlap for anterior neurons while having a separate posterior crista distinct from unique neurons. Tracing in vestibular fibers demonstrates the individual neuronal central projections and forms after the trigeminal fibers begin developing. Fiber projections to the entry point of first vestibular fibers and seem to proliferate after the trigeminal starts in mammals and fish. Interestingly, central projections of vestibular ganglia can navigate after xenoplastic transplanting (mice onto chicken) indicating an overlap of fibers consistent with three ear transplantations in frogs (Elliott and Straka, 2022; Straka et al., 2022).

Early development shows a clear distinction between hagfish, lamprey and gnathostomes based on distinct development: Two distinct vestibular ganglia, an anterior and a posterior branch, are unique to hagfish whereas lampreys and gnathostomes have incomplete segregation of posterior and anterior segregation in a contiguous population of neurons. Central projections to the brainstem show a common projection of a common saccule/utricle in lampreys, hagfish, and gnathostomes. Details of the cristae of hagfish is a circle in the torus and suggest an unusual 'utricle/saccule' projection with a partial overlap in hagfish. In contrast, the lamprey have an organization of a common ganglion that has distinct innervation of three 'cristae' (two anterior and posterior canals, and a third 'dorsal crista') and a single partially separate 'utricle/saccule' (Fritzsch, 1998). Details show unusual projections of the three cristae: the anterior projection is primarily to the posterior caudal projection compared to the posterior projection strongly affects the anterior projection (Fig. 11).

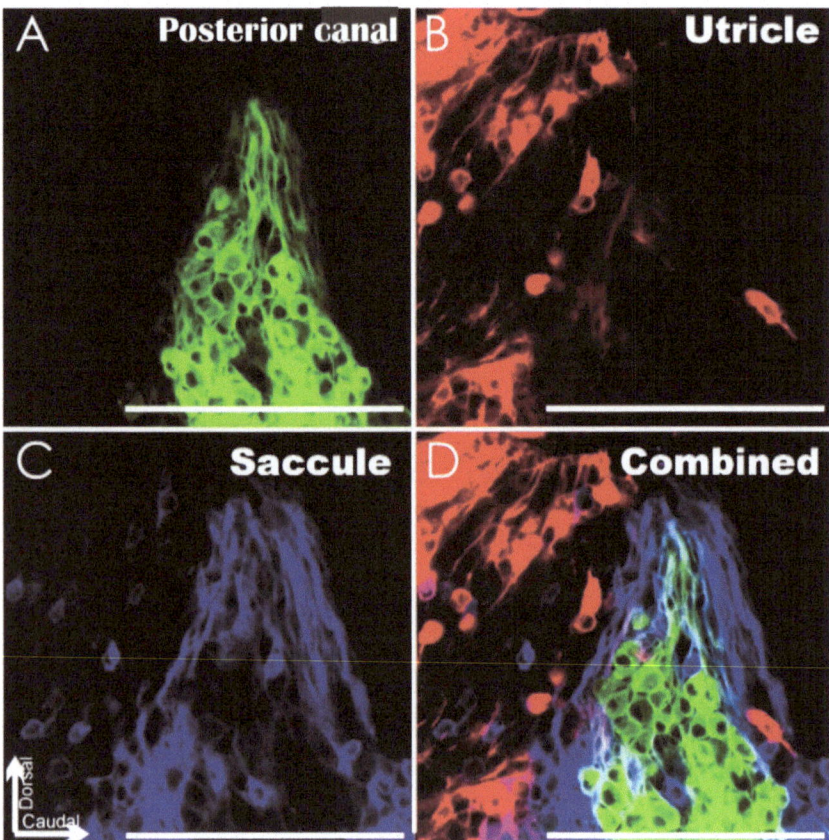

Figure 11: Triple label of vestibular neurons shows that each neuron is unique, labeled by the distinct posterior canal (A), utricle (B), and saccule (C). Despite a distinct label from every three vestibular nuclei, they show incomplete segregation of a closed bundle of posterior canal cristae whereas the utricle and saccule are widespread from those neurons, obviously demonstrated in the combined imaging (D). Bar indicates 100 μm.

In contrast, the unusual projections of the dorsal 'crista' is unique with respect to the central projection, displaying no similarities to the horizontal crista of gnathostomes (Fritzsch et al., 2023a). Details of selective central projections suggest that they are, at the most, incomplete segregation which likely projects overlappingly to all cristae and gravity receptors. Lampreys have different sizes of neurons with few, large neurons that have unclear projections regarding a separate projection from afferents.

In contrast to distinct features of cyclostomes, all gnathostomes have common, contiguous neurons that innervate three canal cristae (anterior, horizontal, posterior crista) to perceive angular information. At least two (utricle, saccule) and up to four 'gravistatic' reception (lagena, neglected papilla) are known with the complete set of a total of nine distinct innervations among amphibians. Details of a few connections demonstrate the incomplete segregation of different neurons forming migrated and intercalated sensory neurons in certain fishes, amphibians, and mammals. A detailed analysis showed an overlap and partial segregation of the anterior crista, horizontal crista, utricle, and saccule that forms a nearly separate posterior crista and nuclei (Fig. 12). Compared to other vestibular neurons, the posterior crista is unique and can have a compact set of neurons that distinguished from the wide distribution of the saccule, utricle, horizontal and anterior cristae. A plethora of branchlets from different neurons can be identified and shows incomplete segregation of the saccule from different branches to almost all vertebrates.

Development of earliest projections have been traced in mice, frogs and lampreys and show the original and earliest vestibular neurons project as first fibers to reach the posterior canal crista (Fig. 12). Detailed adult and development of frogs fit the incomplete segregation of distinct projections. Descriptions from early to postnatal mouse projections clarified the incomplete segregation of only two neurons, the saccule, and the posterior crista. The projection in double color showed incomplete segregation of these fibers but also identified distinct projections none-overlappingly with the two analyzed projections. Sophisticated central projections of five to seven projections require further work using multicolor labeling to detail the sequence of vestibular neurons and how they project distinct and overlapping projections beyond partial descriptions.

An additional projection to the cerebellum exists in some vestibular input but has a projection to two areas of the cerebellum, the nodulus and uvula, in mammals. Based on earlier work, it suggests a partial segregation and shows a complicated partial overlapping. A prominent projection to the nodulus and uvula comes from the posterior canal but receives only fibers from the saccule and uvula. Retrograde labeling of either terminal shows indeed of the canal cristae or are labeled nearly exclusively from the nodulus, whereas the saccule and utricle have a profound projection from the uvula. Details suggest projections develop early and are innervated by

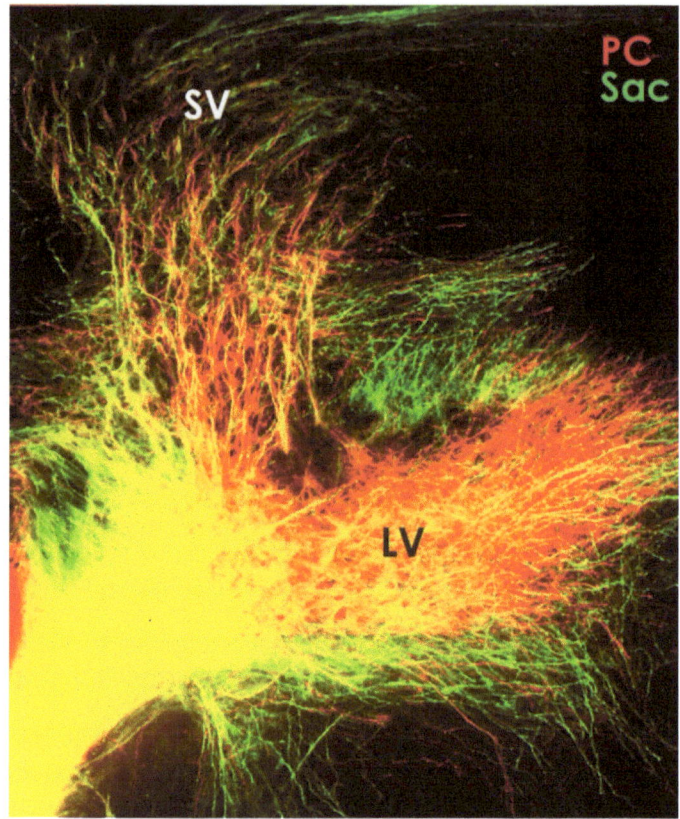

Figure 12: This image shows a coronal section labeled from two nuclei after an injection to the saccule (green) and posterior canal crista (red). Note that the overlap is in the root (yellow) that shows incomplete segregation to the lateral vestibular nucleus (LV) and superior vestibular nucleus (SV). Taken from (Maklad and Fritzsch, 2003).

a part of the saccule and utricle, highlighting half of each terminal of the utricle and saccule. In contrast, additional backfilled fibers from posterior projections highlight complex segregation from discrete projections. Additional information shows the unusual connection to unipolar brush cells and their specific input. More recent details show projections of vestibular branches that are reduced by branches but form a completely separate branch for *Npr2* null mice.

Neuronal development depends on *Neurog1* in mice (around E9), and follows a similar pattern in amphibians, fish, and cyclostomes. Upstream, the neuronal definition depends on *Eya1/Six1* and *Sox2* expression but interacts with neuronal development for normal neuronal development, and affects the horizontal canal by *Lmx1a*, *Foxg1*, *Otx1*, and *n-Myc* (Elliott et al., 2021c). Downstream, neurons are dependent on *Neurod1, Nhlh1, Nhlh2, Tbx1-3, Pou4f1, and Emx2*, among others, followed by the expression of neurotrophin receptors (*TrkB* and to a lesser extent in mammals, *TrkC*). Loss of upstream, *Neurog1*, and both *TrkB/C* results in complete formation or complete early deletion. In contrast, *Neuord1* and other downstream genes resulted in various incomplete losses indicating a vestibular ganglion dependency using reduced vestibular nuclei input. Notably, the absence of vestibular hair cells can develop and project vestibular neurons to the vestibular nuclei normal and indicates an independent projection and shows peripheral sensory epithelia suggesting an independent vestibular system without hair cells. Recent analysis indicates segregation from vestibular is positive for *Tlx3* compared to cochlear neurons that are negative for *Tlx3*. Further segregation of vestibular neurons depends on *Sall3* and *Gata3* which suggests two vestibular neurons, defined as type 1 and type 2 (Sun et al., 2022).

In summary, the neurons are initially formed without hair cell growth. An incomplete and overlapping central projection of all vestibular nuclei has been demonstrated in tetrapods. This incomplete segregation of gravistatic and angular inputs integrates into a single projection that distinguishes it from other central projections from trigeminal/taste ventral and lateral line to the intermediate nucleus dorsal to vestibular nuclei. *Sall3* may be segregation between vestibular neurons of two types. The complexity of incomplete vestibular nuclei can nevertheless have distinct projections to hair cells and have incomplete segregation of central nuclei. Details will be discussed of the peripheral innervation that interacts with hair cells to develop specific and partial innervation of distinct vestibular nuclei to reach specific input.

2.1.2. The vestibular nucleus: Origin and development

A central projection of vestibular ganglion neurons concerns the origin of vestibular nuclei: The continuation of the trigeminal afferents and nuclei can be viewed as a continuation of the trigeminal and other brainstem nuclei. In contrast, the dorsal organization of the trigeminal/spinal cord is different vestibular nuclei, unrelated by any spinal cord, and belongs exclusively to rhombomeres and depend on the choroid plexus. Added to this unique origin, vestibular nuclei are bound by the trigeminal projections ventrally and capped by a lateral line (and electroreceptors in certain vertebrates) that will be replaced by an auditory nuclei among tetrapods. How can a novel set of fibers and receiving their appropriate vestibular nuclei from all vertebrates is in contrast to other chordates? What we know of is a combination of placodes to induce first vestibular innervation by *Neurog1* and shows a reduction of *Neurog1* neurons in the brainstem that are reduced to a transient expression of *Neurog1*, except for the most caudal part. An additional unique feature is formed in *Phox2b* and a split of *Ptf1a* into an additional single *Ptf1a* to add vestibular nuclei. Combined with the unique origin of vestibular nuclei and other aspects around visceral nuclei makes it a unique target for their evolution and development (Glover and Fritzsch, 2022).

The 'vestibular nucleus' is composed of 4 distinct nuclei, the anterior (or superior), lateral, medial, and inferior (or caudal) vestibulospinal tract (AVST, LVST, MVST, and IVST), and is part of the vestibular nuclei in mammals, frogs, fish and lampreys (Fig. 13). The central projection shows incomplete segregation that also has unique inputs from different endorgans. An additional tangential nucleus is known in chicken. Evidence suggests a cross interaction of specific rhombomeres that may depend on

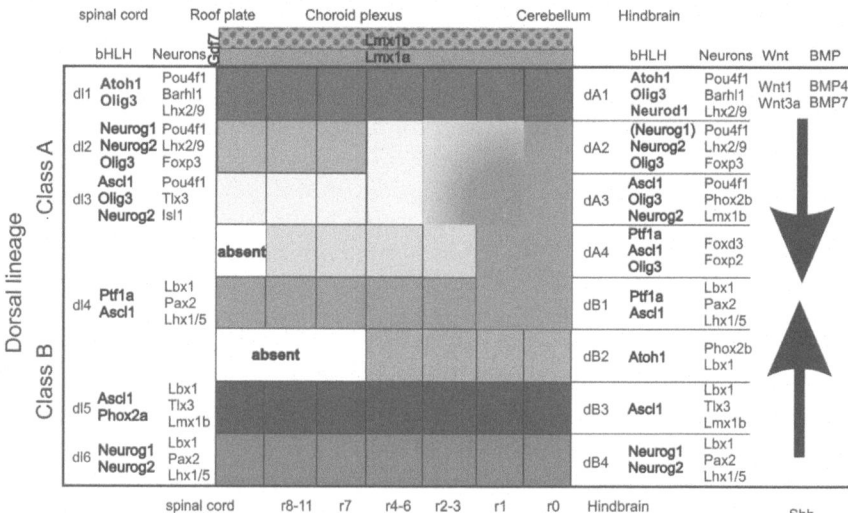

Figure 13: Earliest gene expression of bHLH genes of *Atoh1*, *Neurog1/Neurog2*, *Ascl1*, *Olig3* and *Ptf1a* define the class A and B in dorso-ventral progression in the brainstem (dA1-dB4; right) and spinal cord (dl1-dl6; left). This scheme is changed in the rhombomeres, adding new gene expressions, and deleting others in complicated, sequential changes. For example, dA1, dB1, dB3, and dB4 are present throughout r2-r7 in early expression domains. Noticeable gaps are in dA2 (absent in r2-r6) and dA3 (r2, r3). Interestingly, dA4 and dB2 are present in r2-r6/7 but it is absent in the spinal cord, suggesting a unique addition of the vestibular nucleus of dB2. All early progenitor domains are all positive for *Olig3* (dA1-dA4) and show a progression (top to bottom) of *Atoh1* (dA1), *Neurog1* (dA2, dB4), and *Ascl1* (dA3-dB3). In addition, dA2-dA4/dB1 and dB4 are expressing *Neurog2* and *Ptf1a* (dA4, dB1). Early progenitors can be characterized by homeodomains such as *Pax3* (dA1-dB4), *Pax7* (dA2-dB4), and *Pax6* (dA3-dB4). Interestingly, dB2 is unique by the expression of *Phox2b* in dB2 where it is clear in *Hoxb1* mediated loss of LVST, including *Phox2b*. Vestibular nuclei within a rhombomere background help develop into different areas of the vestibular and adjacent ears. Modified after (Elliott et al., 2021c).

a specific nucleus for a given vestibular nucleus. Beyond a rudimentary understanding of the different vestibular nuclei, it remains to be seen how distinct nuclei develop from distinct dorso-ventral genetic differentiation depending on dorso-ventral patterning driven by *Shh/Bmp/Wnt* gene interaction (Fig. 13). While certain dorsal expressing genes are important for dorsal developing of the brainstem (*Lmx1a/b, Gdf7*) it remains unclear how deeply different dorsal aspects are affected and defects the two vestibular related origins, *Phox2b* and *Ptf1a*.

A clear progression of vestibular nuclei, starting around E10 in mice (E12 in rats) is with the LVST and adds later stages of both anterior and posterior nuclei to the central entry of rhombomere 4 (Fig. 13; Table 1). The first vestibular neuron forms first in mice and nearly exclusively belongs to rhombomere 4 and require *Hoxb1*. A cluster of at least 4 transcription factors defines the dB2 longitudinal column and is expressing *Phox2b, Lbx1*, and *Atoh1*. Additional genes that complement various rostro-caudal extend of vestibular nuclei belong to *Esrg* and *Evx2* and contribute some additional transcription factors are *Onecut1, Maf2, Pou3f1, Lmx1b* and, *Tlx3* in adjacent dA3-dB3. For example, *Lbx1/Tlx3* defines the solitary tract compared to the adjacent vestibular nuclei that are identified by *Lbx1* and *Phox2b* in early domains. Evidence suggests a role for *Ptf1a* to define different populations and can result in their specific differentiation that

Table 1. Distinct bHLH, homeobox, and other transcription factors revealed

Progenitor domain	bHLH TF's	Homeobox domain TF's	Progenitor neuronal transcription factors
dA1	Olig3, Atoh1, Neurod1	Pax3, Msx1	Pou4f1, Barhl1, Lhx2/9, Evx1
dA2	Olig3, Neurog1, Neurog2	Pax3/7, Msx1	Pou4f1, Lhx1/5, Foxd3, Foxp2
dA3	Olig3, Ascl1, Neurog2	Pax3/6/7, Gsx2	Pou4f1, Tlx3, Prrxl1, Isl1, Phox2b, Lmx1b
dA4	Olig3, Ascl1, Neurog2, Ptf1a		Foxd3, Foxp2
dB1	Ascl1, Neurog2, Ptf1a	Pax3/6/7, Gsx1/2	Lbx1, Pax2, Lhx1/5
dB2		Phox2b	Lbx1, Phox2b, Atoh1
dB3	Ascl1	Pax3/6/7, Gsx1/2, Dbx2	Lbx1, Tlx3, Lmx1b, Prrxl1, Pou4f1
dB4	Neurog1, Neurog2	Pax3/6/7, Dbx2	Lbx1, Pax2, Lhx1/5, Wt1, bHLHb5, Dmrt3

Modified after (Fritzsch, 2021; Hernandez-Miranda et al., 2017)

overlaps with dA4/dB1. Obviously, the length of expression from E10 to adults may require some additional transcriptions to fine tune the details (Glover and Fritzsch, 2022).

The expression of at least *Phox2b* and *Lbx1* is unique along the entire ventral column and indicates a unique origin of at least these transcription factors, unparalleled to the column of dB2 nuclei that are not found in the spinal cord (Fig. 13; Table 1). Combining the early vestibular origin suggests an earliest r4 origin ahead of all other vestibular nuclei in terms of proliferation, indicating an earlier detailed variation from various genes distinct from other spinal cords. Superimposed is the choroid plexus that defines *Lmx1a/b* as well as the *Gdf7*, *Atoh1*, *Ptf1a*, *Wnt1*, and *Wnt3a*. In the absence of both *Lmx1a/b* (Fig. 13) ceases formation of *Gdf7*, *Atoh1* (dA1, dB2), *Ptf1a* (dA4, dB1), and reduced *Bmp's* and *Wnt1/Wnt3*, but it is unclear how far the dorsal definition is affected by other genes, notably *Phox2b* (dA3, dB2) and *Tlx3* (dB3). It should be pointed out to suggest a unique expression of *Phox2b* that is selectively expressed in rhombomeres whereas the spinal cord is expressed of *Phox2a* that lacks 'rhombomeres'. An interesting pattern of central projection develops in an unusual, bilateral interaction of vestibular fibers in the absence of the choroid plexus: instead of segregated from the choroid plexus, left and right projections are interacting across the 'roof plate' of vestibular, taste, and trigeminal fibers, crisscrossing with left and right input (Chizhikov et al., 2021).

Description of rhombomeres is clear, but it is unclear how the different rhombomeres are interacting with other transcription factors. A longitudinal connection is partially present for chickens, mice, and frogs but requires additional research to detail the nuclei with the different outputs to the spinal cord and motor neurons to impact the output. Detailed input is followed by the three major vestibulo-spinal projection neuron groups, the LVST (lateral vestibulospinal tract group), the iMVST (ipsilateral medial vestibulo-spinal tract group), and the cMVST (contralateral medial vestibulo-spinal tract group). Further caudal description is unclear concerning the caudal vestibulo-spinal tract group (Glover and Fritzsch, 2022). In addition, four main vestibulo-ocular projection neuron groups convey vestibular signals to all the extraocular motoneurons except those in the abducens nucleus. Finally, the vestibulo-cerebellar projection neurons belong to gnathostomes but are absent in the cyclostomes that have no clear identification beyond the transmigrated trochlear neurons, leaving the migrated 'trochlear nuclei' is a major 'cerebellar nucleus' in lampreys and is absent in hagfish.

Most studies are incomplete with respect to the different endorgans projections to distinct vestibular nuclei and how the output combines with inputs, helping to define the various central projections to the spinal cord, cerebellum, and ocular motor neurons have never been studied.

Ideally, a complete distinct labeling from each endorgan in combination with targeted projections to the contralateral vestibular complex as well as the ocular motor neurons, spinal cord, and cerebellum, would be needed to fully correlate the connections with the advanced level of vestibular output, in frogs and mammals. Details will be needed to fill the similarities and distinctions of vestibular input-output relationships for which we are starting on the two distinct cyclostomes: the hagfish without a cerebellum and a lack of ocular motor neurons and compared to the lamprey that the unique changes and distribution of a 'cerebellum' filled with trochlear motor neurons and the unique 'octaval' output from three major populations connected with giant interneurons, the Muller and Mauthner cell neurons (Chagnaud et al., 2017).

The apparent longitudinal expression of *Atoh1* from the spinal cord to the cerebellum is defined by lateral line, electroreception, and auditory nuclei that are in continuation with the spinal cord, indicating a discontinuity between the dorsal dA1 domain of the brainstem that is not in continuity with the equivalent of *Atoh1* expression as well as *Neurog1*. The distinct difference of dA1 for all three receiving projections is distinct and contrasts with the composition of at least three longitudinal expressions (dA4, dB1, and dB2) to compile the vestibular nuclei and is distinct from afferent projections compared to lateral line, electroreception, and auditory projections (Elliott and Straka, 2022; Glover and Fritzsch, 2022; Straka et al., 2022).

In summary, the current work is incomplete in connecting the rhombomeres with the scaffold of vestibular nuclei except for *Phox2b*, *Lbx1*, and *Atoh1* of dB2 and at least two additional longitudinal projections (dA4, dB1) that expand in more dorsal domains extending it beyond dB3 (r4-6) and have an additional expression of *Neurog1* in r7 and the spinal cord. Details selective projections are needed to demonstrate the domains with distinct vestibular, taste and trigeminal projections (Fritzsch et al., 2022). While it is a dominant player for *Lmx1a/b* expression of the rhombencephalon, details are needed to define the loss of vestibular neurons with altered inner ear projections. What is required to establish the domains among lampreys in comparison to frogs and mice to define the vestibular nucleus compared to other genes.

2.1.3. Vestibular hair cells

Hair cells are well described in various vertebrates and depend on *Atoh1*, following is the initial description of two types of hair cells, Type 1 and Type 2, in amniotes (Lysakowski, 2021). Upstream and downstream genes have been identified and are needed to express normal hair cell development in the vestibular hair cells. A set of upstream genes depend on earliest expression are *Foxi3* and *Fgf3/10* to induce placode formation and requires *Brg, Eya1/Six1, Pax2/8, Sox2, Foxg1, Neurog1,* n-*Myc* and *Lmx1a, BMP4, Otx1, Emx2, Srrm4, Tbx1-3, Tlx3* and other genes for hair cells (Elliott et al., 2021c; Zine and Fritzsch, 2023). Loss of all hair cells has documented in *Foxi3, Pax2/8* and *Eya1/Six1* and all hair cells depend on *Sox2* and *Atoh1*. Interestingly there are certain losses of hair cells such as *Foxg1* (lack of the horizontal cristae, also deleted in n-*Myc* null) and cause reduction, fusion, and expansion for *Lmx1a* deletions and fuses the horizontal connections with the utricle in *Otx1* mice. *Tbx1-3* are affecting the vestibular hair cells but requires a detailed description (Kaiser et al., 2021). An incomplete loss of *Neurog1* deletions leads to near complete loss of the saccule and hair cells but are short of the utricle and all three cristae which are reduced in *Gata3* and *Prox1* deletions.

Downstream of *Atoh1* is *Pou4f3, Gfi1, Barhl1* and others. An interesting vestibular loss is due to the absence of *Srrm4* which will be 100% loss in *Srrm3/4* double null mice that is interaction for REST, resulting in all vestibular hair cell development. Likewise, in *Emx2* deletion that affects some, but not all hair cells. Of interest is the delayed information of conditional deletions that show a partial and incomplete deletion in combination with a delayed deletion of *Atoh1*-cre; *Atoh1*f/f. Likewise, the effect of loss of *Lmx1a* causes an organized vestibular projection and reduces some areas while it expands other hair cells (Elliott et al., 2021c). Additional complex interaction is provided with the *miR-183, 182* and, *96* (Fig. 14) that requires all hair cells and depends on the dicer gene, and can result in nearly all vestibular neurons in a *Dicer* conditional deletion in the mouse.

Upstream of *Atoh1* provides crosstalk to other bHLH genes such as *Neurog1* and *Neurod1* that are immediately upstream of *Atoh1*. All hair cells depend on *Sox2* and other *Sox* genes and depend on *Brg-BAF* chromatin remodeling and *Eya1/Six1* which are essential steps in the development of hair cells (Xu et al., 2021). While multiple genes are analyzed in particular in mice, it is unclear how the different hair cells develop. The overall simple appearance of a single torus that has two cristae and a single common macular has demonstrated to be a derived development and is not a primitive hagfish (Fig. 14). Obviously, the unusual distribution of a single torus that lacks a cupula and forms a single canal throughout the ear is best demonstrated when using specific staining such as miR-183. Likewise, the common macula lacks any distinct divisions that are dependent on *Emx2* in bony fish and mice.

Genetic data suggest that cyclostomes are closely related in lamprey and hagfish but also show that lampreys are an earlier branch of lampreys (Fritzsch et al., 2023a) while hagfish are derived. While a similar unified common macula that shows at the most lacks some similarity to hagfish, its detailed organization is different compared to two-three distinct patches with a different organization of hair cells, the utricle, saccule, and lagena (variably present among jawed vertebrates).

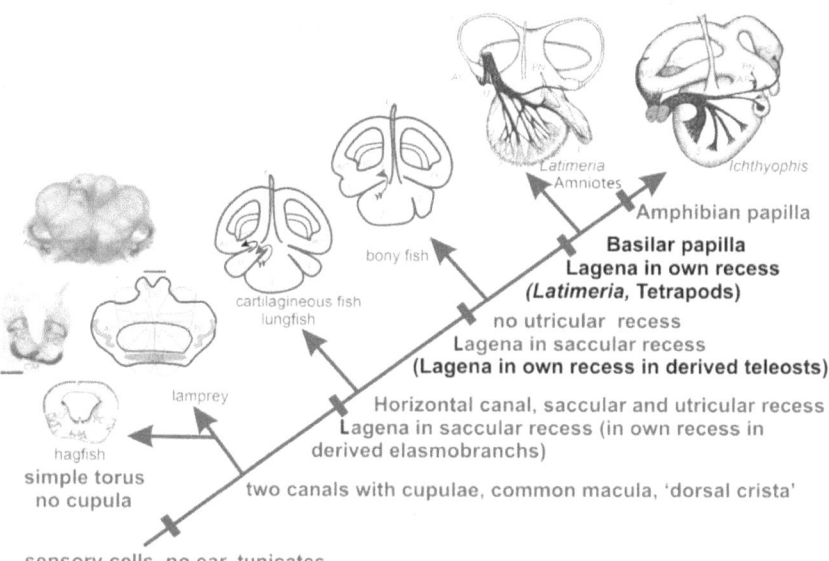

Figure 14: The adult vestibular organs can develop from at least two cristae (anterior and posterior crista) and a common macula, exemplified in hagfish. The unusual organization of a torus of cristae in hagfish (shown with miR-183 ISH) is a derived feature, somewhat exclusive of lampreys that have a unique pattern of hair cells in the cristae. An additional dorsal crista (DC) in lampreys is a unique feature that might evolve into the lateral or horizontal canal (HC) in all jawed fishes including tetrapods. Note that the subdivisions of the common crus show in tetrapods and bony fish that display additional separation of the utricle (U) in cartilaginous fish and lungfish compared to tetrapods and bony fish. The common macula is revealed by the incomplete segregation of hagfish and lampreys. In contrast, at least two distinct recesses, the utricle and saccule (S) are present among jawed vertebrates. An additional third gravistatic forms in derived gnathostomes, the lagena (L), that disappears in derived mammalian through losses. Note that the number of additional features is the basilar papilla (BP) among almost all tetrapods and *Latimeria* whereas the unique formation of the amphibian papilla that develops through division of the amphibian/neglecta into a total of 9 sensory epithelia, the largest distinct separations among caecilians. Modified after (Fritzsch and Beisel, 2004; Fritzsch and Wake, 1988; Pierce et al., 2008).

All hair cells follow the same principle of a staircase organization of longer stereovilli (stereocilia) that is associated with a much longer kinocilium in all vertebrate vestibular hair cells. This basic principle comes about through the interaction of different genes that define the organization of the asymmetric distribution among hair cells and drives the reorganization of different types of hair cells. Cristae can be separated by hair cells based on the diameter of stereocilia: thick stereocilia and thinner stereocilia are mixed among the tetrapod whereas hair cells are identical among jawed vertebrates. In contrast, the organization among lampreys has differed in length into three categories, short stereocilia, intermediate stereocilia, and tall stereocilia. In contrast to the sister group of jawless vertebrates, the hagfish have uniform and extremely long stereovilli that are not developing as a cupula. Combined, lampreys and hagfish are unique from the jawed vertebrate that has little differences beyond the two types of hair cells in tetrapods (Elliott and Straka, 2022).

Vestibular hair cells consist of two to three maculae for gravistatic reception and two to three canal cristae for angular reception. Polarity depends on function, but the distribution of hair cells differs. Only maculae have opposing maculae (Ji et al., 2022), whereas canal cristae are uniform in their polarity. Canal cristae are also present in most auditory hair cells. All vertebrate hair cells have stereocilia organized in a staircase pattern, displaying distinct apical polarities for stimuli to open mechanoelectrical transduction channels (METs) by tip links using *Pcdh15* and *Cdh23*, permitting endolymphatic potassium to enter the HCs and change their resting potential. The mammalian mechanosensory channel is, in part, formed by the transmembrane proteins *Tmc1* and *Tmc2*. Other interactions are known, but these interactions require additional work for the MET formation. A unique formation of vertebrate hair cells is found in the *Tmc1/2* single gene in cyclostomes. *Tmc1/2* is separated from the closely related gene, *Tmc3*. However, the function of *Tmc3* is unclear in nearly all animals, including basic animals, for which there is no information regarding its function. Additional information is needed for *TMEM16* and *Piezo* to fully understand the MET.

The otoconia bearing gravistatic receptors differ among jawless and jawed vertebrates: all jawed vertebrates have at least two distinct maculae (utricle and saccule) that form two opposing polarities. In contrast, the common maculae of hagfish and lampreys show minor differences that never develop opposing polarities. Evidence of *Emx2* is essential for the two opposing polarities in bony fish and mice. Planar cell polarity (PCP) genes depend on *Frizzled, Prickle, Disheveled, Van Gogh, Diego,* and *Flamingo* for normal development. Polarization depends on *Emx2*, which eliminates the contralateral organization in the utricle by converting it into a single polarity (Ji et al., 2022).

In addition, retinoic acid (RA) sets up various gradients. Saccule and lagena have different polarities. Instead of polarizing each other again in the utricle, they flip to organize in the saccule and lagena. A distinct pattern of the utricle and saccule has a separate innervation from the cerebellum to reach one polarity and receive a descending branch of the caudal vestibular neurons to end up in a different innervation.

In summary, the vestibular neurons depend on *Neurog1*. The central vestibular nucleus has several genes that play a role but have not yet determined the formation except for the role of r4 that drives the lateral vestibular neurons, the Deiters' cells connecting the spinal cord. In contrast, all vestibular hair cells depend on *Atoh1* which develops into two types of amniotes.

2.2. Lateral line evolution seemingly coincides with vestibular development

The lateral line is a sensory system that allows vertebrates to sense the pattern of water flowing over their body surface via mechanosensory organs, called neuromasts, which are distributed in a characteristic pattern over the body surface. The evolution of the lateral line system is part of the lateral line component of craniates. In contrast, all amniotes have no development of a mechanosensory lateral line system beyond some preliminary precursors. A unique population is found in several gnathostomes, the spiracular organs, which equates to the lateral line system (O'Neill et al., 2012).

The mechanosensory cells are innervated by sensory neurons and are unique among tunicates and craniates that connect secondary sensory cells and reach the brain connected by a unique set of neurons. Among chordates, tunicates have true cilia innervated by secondary sensory neurons that express *Atoh* to develop sensory cells. Connections will develop with *Neurog* to form 'ganglia' adjacent to the spinal cord. In contrast, the lancelet only has fibers that project to the CNS directly without showing the formation of 'ganglion neurons' that lack *Atoh* and have a short expression of *Neurog* expressed inside the brain. Multiple 'mechanosensory' cells are known among jellyfish that indicate the evolution of a stepwise transformation in sensory cells, starting with choanoflagellate. The evolution shows the molecular basis of TMCs for the likely mechanosensory hair cell transduction that shows the multiplication of TMCs in chordates.

Vertebrate lateral line sensory evolution shows distinct neuronal projections that innervate different sensory hair cells, comparable to the vestibular hair cells, and connects them centrally to the alar plate of the hindbrain (Fig. 15). Proper sensory perception requires the interaction of a unique set of sensory neurons projecting into distinct lateral line central nuclei in vertebrates, and receive sensory hair cells and the unique efferent innervation (Engelmann and Fritzsch, 2022):

1. Sensory lateral line neurons develop from anterior and posterior neurons comparable to VII/IX neurons that connect sensory hair cells at the periphery and alar plate nuclei centrally.
2. Hair cells are innervated from anterior and posterior *Neurog1* ganglia that innervate *Atoh1* positive hair cells.

3. Nuclei in the alar plate of the hindbrain require *Atoh1* genes and their development and maturation involve neuronal inputs from distinct targets.

The lateral line nuclei in the alar plate of the hindbrain, the mechanosensory hair cells at the periphery, and the sensory neurons that connect them are referred to as the 'octavolateralis' system. Sensory neurons depend upon the expression of two basic helix loop helix (bHLH) transcription factors, *Neurog1* and *Neurod1*. Across sensory lateral line systems, there is a difference in how sensory information is relayed to the brain. There is a clear developmental progression of the lateral line sensory cells as well as a sequential progression of sensory neurons that project from these hair cells to distinct rhombomere nuclei with respect to other sensory systems. Furthermore, there is a delay of peripheral sensory target innervation after sensory neurons develop and become connected to the different rhombomere nuclei targets. The sequence of specific sensory cells is unique for a set of chordates that evolved later in the development of craniates (cyclostomes and gnathostomes) as their shared central projections may differ among tunicates that is lacking a lateral line innervation in the lancelets.

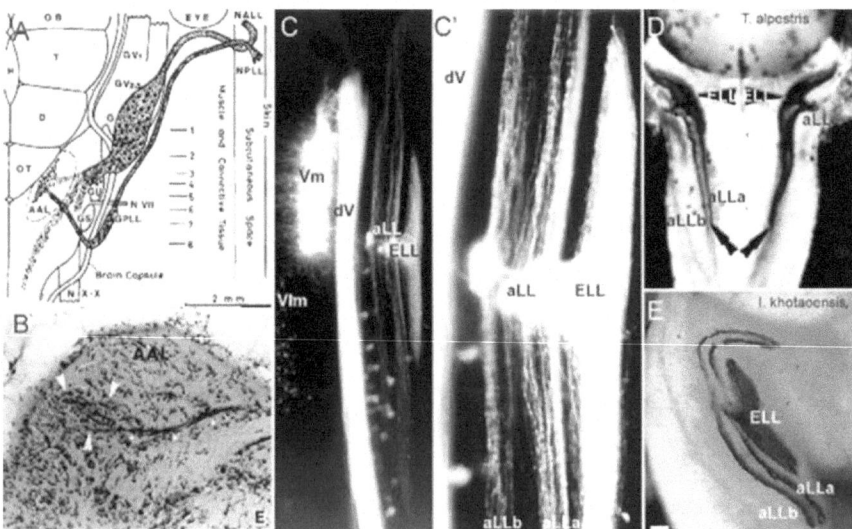

Figure 15: Lateral line projections (LL) are present in hagfish (A, B), lampreys (C, C'), salamanders (D) and caecilians (E). A discrete central projection forms two branches, presenting the two hair cell polarities that overlap in hagfish. Note that the presence of electroreceptors (ELL) is shown in lampreys, salamanders and caecilians that are absent in hagfish. Modified from (Elliott and Fritzsch, 2020).

2.2.1. Lateral line sensory neurons: Define the molecular base of neurosensory formation from the lateral line placode

1.4 million animals evolved from only 263 choanoflagellates species of which only a small group of 44,000 vertebrate species have at least some lateral line systems (Fig. 5). Those that have lost the lateral line system, including all amniotes and derived amphibians, may, at best, have incomplete lateral line development that forms a unique population in gnathostomes, the spiracular, or paratympanic organ. A derived hagfish has a lateral line hair cell without opposing polarity. A unique basic group of hagfish is virtually unknown with respect to the lateral line that remains to be analyzed. The parallel evolution of a 'lateral line' is present in cephalopods and is driven by distinct multiple kinocilia defining a parallel evolution. In contrast, lampreys have opposing hair cells that form distinct groups of neuromasts, comparable to all gnathostomes, and all vertebrates have two opposing hair cells innervated by at least two distinct nerve fibers (Figs. 5, 15).

The lateral line system forms an anterior (aLL) and a posterior (pLL) system (Fig. 15). The aLL innervates neuromasts of the head and the pLL innervates neuromasts of the trunk/tail. In general, lateral line neurons project along the three major trigeminal branches (ophthalmic, maxillary, and mandibular branches) starting from the facial nerve to provide aLL nerves. A separate pLL starts from the glossopharyngeal nerve to branch into several branches. aLL and pLL projections are discrete from facial and glossopharyngeal nerve fibers, enter the hindbrain near r4 and r6, respectively, and show independent projections to distinct branches of lateral line central nuclei (Fig. 15).

Most fish continuously produce neurons, adding sensory neurons, sensory nuclei, and sensory hair cells for a long time, showing a completely different progression compared to most mammals. Sensory neurons, hair cells, and nuclei critically depend on bHLH genes for their differentiation and development. Historically, the first insights into four early genes were found almost simultaneously in mice:

1. *Sox2* expression provides the precursors to initiate
2. *Neurog1* has a role in sensory neuron differentiation, including lateral line afferents.
3. *Neurod1* acts downstream of *Neurog1* and is needed by neurons and hair cells.
4. *Atoh1* is essential for lateral line hair cells and lateral line nuclei.

Assuming what is true for the mammalian ear development is also true for all other neurosensory and rhombomere nuclei among other vertebrates. We distinguish three interrelated issues of sensory formation, 'hair cell' development, and hindbrain alar plates, consistent with the molecular biology of different vertebrates. I will describe the development of lateral line neurons followed by the analysis of common and unusual loss and gain of lateral line nuclei (intermediate nuclei) which is followed by 'hair cells' (lateral line 'neuromasts' of mechanosensory hair cells).

The lateral line has been described in larval lampreys and their placodal development has been detailed in several presentations. Evidence suggests an early development of lateral line innervation (Engelmann and Fritzsch, 2022). Little information exists in chondrichthyes and teleosts, except in three areas where we have gained some insight into the lateral line of sturgeons, knife fish, and zebrafish. The work of Baker and collaborators (Baker, 2019) has provided an expansion of our understanding of the lateral line and has provided limited evidence for the central projections beyond establishing the different targets. No detailed central projection tracing has been done yet during development in most gnathostomes. Interesting data have been generated in the developing knife fish that suggests sensory neurons may induce the formation of the mechanosensory cells at the periphery. By far the best study of innervation is in zebrafish with respect to lateral line development. Most importantly, the sequential central projection following trigeminal > inner ear > lateral line projection was provided in parallel to the sequence of lateral line development.

Lateral line projections are found in nearly all craniates (Fig. 15). Cyclostomes have unmyelinated nerve fibers. These nerve fibers are myelinated in gnathostomes. The lateral line fibers are separated typically into two distinct fascicles, but it is less clearly separated into fascicles in other species. Lateral line innervation of hair cells varies between minimally two afferents but can have multiple afferents in larger target areas. Tracing independent branches is difficult due to the afferents from different hair cell polarities that run together in some areas of a given lateral line nerve. Interestingly, some posterior lateral line afferents of lampreys project to the 'cerebellum,' while lateral line afferents of other species do not extend beyond r2 with exceptions. Despite the exuberant expansion of most derived gnathostomes, lateral line fibers generally project comparable to other species with respect to the various branches of aLL and pLL fibers. At the periphery, afferents innervate lateral line hair cells at 'bar' or 'ribbon' synapses (Engelmann and Fritzsch, 2022).

Lateral line afferents follow the path of hair cell precursors that is among the first tracing of amphibian lateral line projections to the target. The initial findings of Herrick (Herrick, 1948) were used extensively by others to study primarily the posterior lateral line innervation. In contrast, the elongation of the lateral line system is following as the neuromasts instead of following the posterior latera line system. Unlike the lateral line efferent innervation of gnathostomes, neither lamprey nor hagfish have efferents. Efferent neurons terminate on acetylcholine nicotinic receptors.

Hair cells express α9 and α10 nicotinic acetylcholine receptors prior to efferent innervation for some hair cells and after efferent innervation for others.

In summary, lateral line innervation is from two branches (aLL and pLL), connecting the peripheral hair cells and the intermediate nucleus with at least two afferents per hair cell. In addition, lateral line hair cells are innervated by a single efferent, except in lampreys and hagfish.

2.2.2. Projections of the central lateral line projections

The lateral line consists of two distinct nerve fibers, the aLL and pLL, and enters the hindbrain at r3-4 (near facial nerve) and r5-6 (near glossopharyngeal nerves), respectively. The reorganization of the inner ear will reorganize the central projections more clearly in the absence of the inner ear. Central projection of lateral line afferents is identical in hagfish, lampreys and most gnathostomes, except certain amphibians and all amniotes. The nuclei in the alar plate where lateral line fibers (aLL, pLL) project to is known as the intermediate nuclei [IN] or medial octavolateralis nuclei [MON]. As previously noted, it is clear that the octavolateral system implies a mix of octavo (vestibular) and the lateral line nerve connections. However, there is no association of the lateral line fibers with 'octaval' fibers from vestibular neurons projecting to the vestibular nuclei. Instead, aLL and pLL fibers enter the hindbrain distinct from vestibular fibers. The boundary of lateral line will depend on certain groups that have lost lateral line innervation in many adult frogs, some caecilians and salamanders, and all amniotes.

Lateral line projections into the intermediate nucleus show a unique feature of interdigitating aLL and pLL fibers in some species, but in other species, these fibers remain segregated. For example, a clear projection of two distinct fibers per each aLL and pLL is traced from the hair cells are known for salamanders and caecilians that have an overlap of all fibers from either central projection in *Xenopus*. A clear specific projection of distinct aLL and pLL fibers are well studied in the zebrafish organization. In addition, lateral line afferent fibers project in two segregated projections either to the Mauthner cell system, where these projections can initiate quick motor behavior response, or to an ascending pathway, where an ordered topographical representation of the sensory surface is formed that mirrors the body surface (Engelmann and Fritzsch, 2022).

The lateral line nuclei of lampreys are differently organized from other lateral line nuclei targets (Fig. 15). In lampreys, the pLL fibers cross r1, whereas these fibers do not project beyond r2 in gnathostomes. This distinction of more rostral projection of the pLL is reminiscent of posterior vestibular fibers projecting rostral, whereas anterior vestibular fibers project predominantly to caudal projections of vestibular targets.

Incidentally, the lampreys have altered visual connections of a dorsal trochlear motoneuron overlapping with lateral line fibers, whereas hagfish have no visual oculomotor innervation. The projections to the anterior lateral-line nerves project to the anterolateral portion and the posterior lateral-line nerve to the caudal portion from a topographic organization. The topography is not precise as the projections of the branches of the anterior and posterior lateral-line nerves.

In summary, the IN forms a common central nucleus for the first order of lateral line nerves (aLL, pLL) that show variable degrees of segregation of afferents. Except for lampreys, lateral line nuclei projections span from r2 to r7, slightly less than vestibular fibers from r1-r8.

2.2.3. Innervating lateral line nuclei in the brainstem

Since described the role of the rhombencephalon development in mammals, it started with novel ideas of different roles of central nuclei (Glover and Fritzsch, 2022). How are these mechanosensory systems organized to connect hair cells to central nuclei via sensory afferent neurons? Molecular information to generate discrete and overlapping projections is needed to provide a common theme of distinct losses and gains across vertebrates. Starting with the molecular background of defining the various genes that could play a role in alar plate nuclei formation, including dorsal relevant genes such as *Lmx1a* and *Lmx1b*, *Wnts*, *Sox2*, *Atoh1*, *Ptf1a*, and micro RNAs.

Specific early expression of genes, such as *Atoh1*, is known in developing dorsal alar plate nuclei, such as in lampreys, bony fish, and mammals. *Atoh1* is expressed in all intermediate nuclei, much like the *Atoh1* function plays a major role in cochlear nuclei in mice. Other genes are regulating for the *Sox2* and *Sox3* and interact with *Lmx1a/b*, which is needed for *the Atoh1* expression, and all may be involved with the earliest lamprey projections.

Recent work in mice showed that deletion of both *Lmx1a* and *Lmx1b* causes a complete loss of *Atoh1* and certain *Wnt's* including *Wnt1* and *Wnt3a*. In *Lmx1a/b* null mice, there is a single dorsal expression of *Wnt3a* instead of two parallel Wnt expressions between the choroid plexuses. The choroid plexus is highly positive for *Lmx1a* and *Lmx1b* and loss of *Lmx1a/b* abolishes *Atoh1* in the dorsal hindbrain of mice. The loss of auditory nuclei in *Lmx1a/b* null mice is likely due to the loss of the downstream gene, *Atoh1*, which is comparable to the loss of nuclei described in *Atoh1* mutants. In the absence of *Atoh1*, and thus central cochlear nuclei, auditory afferents project nearly identical to that of wild type mice. However, in *Lmx1a/b* null mice, central projections are aberrant. Due to the fusion of the dorsal roof plates in *Lmx1a/b* null mice, fibers of the inner ear project bilaterally at the roof plate, showing incomplete segregation of inner ear fibers as they intertwine. With this clear outline, we can compare the early deletions across craniates, including other genes such as *Ptf1a*.

In lamprey and hagfish, some relevant genes are expressed, including *Atoh1* and *Ptf1a*. As described by others in the development of mice, *Atoh1* is expressed dorsally along the spinal cord and hindbrain to reach the cerebellum in mice. Given the similar gene expression in auditory nuclei in mammals that replaces the equivalent nuclei of lampreys, we can assume that loss of *Atoh1* expression will eliminate lateral line nuclei in lamprey. However, lampreys and hagfish have a single *Lmx1a/b* gene, as compared to a separate *Lmx1a* and *Lmx1b* in all gnathostomes. Following the data generated by *Lmx1a/b* double null mice suggests that neither lamprey nor hagfish would develop any lateral line nuclei following the loss of the *Lmx1a/b* gene. There is incomplete data for any additional *Wnt*'s that may be needed in hagfish and lampreys.

Among bony fish, only zebrafish data have provided some clues about different genes in development. First, early gene duplication evolved into an *Atoh1a* and an *Atoh1b* that changes in early expression in the ear and brain. In addition, the ubiquitous early gene expression of *Sox2*, necessary for early proneural bHLH gene expression, shows interesting interactions of *Sox2* and *Sox3* and with downstream genes of *Atoh1*, *Neurod1*, and *Neurog1*. Whether *Lmx1a* and *Lmx1b* play a role in bony fish has not yet been analyzed.

Only recent data are provided for salamanders and frogs and show little expression of *Atoh1*, *Sox2*, and certain other genes but central projections were not analyzed in older amphibians. Several *Lmx1a* and *Lmx1b* genes have been studied for early or very late expression in *Xenopus* but have very few data have been provided. Different genes that could be eliminated and/or misexpressed are needed to fully understand the molecular basis for the brainstem development of amphibians, including salamanders (Engelmann and Fritzsch, 2022).

The intermediate nuclei overlap with *the Atoh1* expression more ventral and likely depend on *Lmx1a/b* expression for their formation, as is the case for auditory nuclei. CRISPR can be used to study the effects of *Lmx1a/b* knockdowns in lampreys, teleosts, and amphibians to correlate with the various dorsal alar plate nuclei and choroid plexus formation. Expression of *Lmx1a/b* can relate to the loss or gain of mechanoreception of different kinds of bony fish, hagfish, lampreys, and amphibians. *Lmx1a/b* and downstream *Atoh1* and *Wnt3a* could explain the alar plate and choroid plexus formation in these species that is lacking a choroid plexus in hagfish.

Dye tracings of various central projections of the salamander, Axolotl, through development showed distinct progression of trigeminal > inner ear > lateral line > electroreception. In fact, different branches of lateral line projection differ between more ventral and more dorsal of the lateral line projections, relating that to the polarity among mammals in the vestibular system. Combined, these data point out a larger polarity projection in lateral line and vestibular projections. Limited work on central projection has occurred in lateral line central development in anurans but suggests a similar delay in various central projections may happen in anurans and caecilians.

Salamanders, caecilians, and anurans have various levels of degeneration of different lateral lines after initial formation in larvae. The lateral line remains during aestivation, only to emerge again to swim in water and engage in reproduction. Transplantation of ears to the trunk of *Xenopus laevis* embryos demonstrated unique interactions between the ear and the lateral line afferents. Lateral line afferents were observed navigating along inner ear afferents projecting into the inner ear and randomly project to the anterior or posterior expansion along lateral line fibers. These data suggest that the segregated projections of afferents develop with a delay to avoid cross-innervation of different sensory systems.

In summary, we are beginning to understand a novel insight driven by newly uncharacterized genes that are essential for all dorsal parts of the hindbrain, including alar plate and choroid plexus formation. The common theme might be these genes are driving the development of the dorsal parts of the alar plate, including the intermediate nucleus and auditory nuclei.

2.2.4. Define hair cells and their polarities to define molecular basis of lateral line hair cells

Nearly all craniates have lateral line hair cells, including lampreys and hagfish, which lacks in amniotes. Hair cells consist of up to several hundred or even several thousand superficial neuromasts. Of all craniate lateral line systems (Fig. 16), lampreys and hagfish are less derived but have the same organization of hair cells and supporting cells. However, further work is needed to describe an evolutionary earlier branch of hagfish. Lateral line hair cells are organized into two opposing polarities. The conversion of a mechanical stimulus into an electrical signal requires a kinocilium and gradually shorter stereocilia (Fig. 16) in a clear polarity to function as mechano-electrical transduction (MET). The tip link connects the stereocilia by threads of *Cdh23* and *Pcdh15* in mammals and uses the same tip links in zebrafish. Additional connections are needed for normal function, including *Tmc1/2, TMIE, Clarin, Myo7,* and *Myo6*. Detailed physiology involving *Vglut3* as the major synaptic charges is dependent on ribeye and other genes.

The mechanosensory transduction connection of kinocilia and stereocilia may not work identically in gnathostomes and

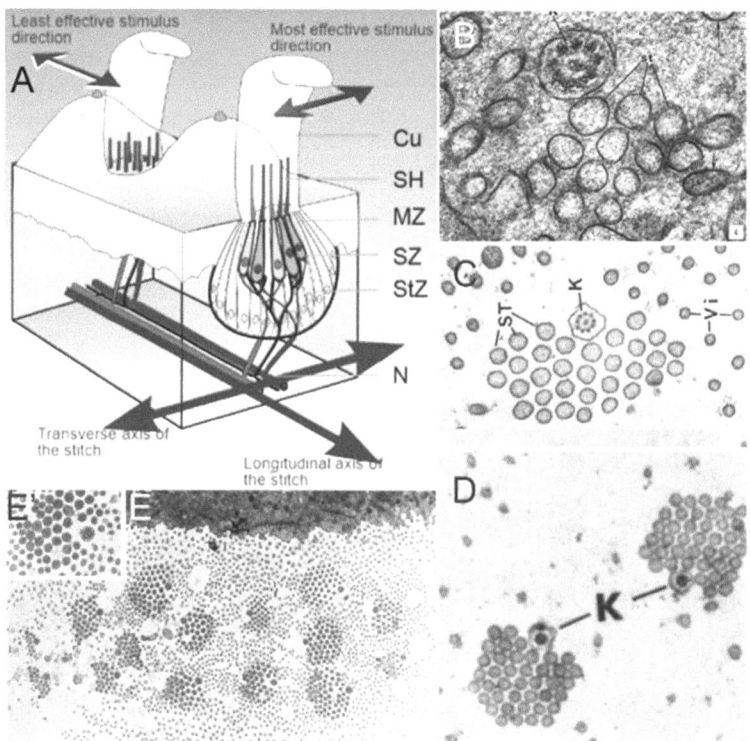

Figure 16: Hair cells are polarized in *Xenopus* (A), 20 stereocilia in lampreys (B), show 30-40 stereocilia in sharks (C) and can have 60 or more stereocilia in eels (D) and caecilians (E, E'). Note that opposing is random (A) or counteracted (D, E) each other by kinocilia (K). Modified after (Elliott and Fritzsch, 2020).

cyclostomes. *Tmc1* and *Tmc2* are the likely protein candidates for the mechanotransduction channel. The absence of *Tmc1*, *Tmc2*, or TMIE disrupts normal stereocilia development. Only a single *Tmc1/2* gene is known for lampreys and hagfish and may relate to the distinct differences in lamprey mechanosensors. Moreover, lamprey lateral line hair cells have few and very short stereocilia. Combined, these features set aside lamprey and hagfish from 'typical' mechanosensory lateral line hair cells (Engelmann and Fritzsch, 2022).

All lateral line-containing gnathostomes have a similar organization of kinocilia and stereocilia (Fig. 5). Gnathostomes have more stereocilia (50-60 stereocilia) compared to lampreys (~18-20 very short stereocilia). Lateral line hair cells show more stereocilia in Chondrichthyes (about 30-40 stereocilia) that are short compared to the very long kinocilia in most lateral line hair cells. Peripheral mechanosensory hair cells form clusters with different numbers and polarity distribution of hair cells.

They may be organized in a random distribution of the two opposing cell orientations or in a counter organization. Most lateral line systems have a distinct distribution among gnathostomes: a superficial lateral line and a deep lateral line that sinks into tubes. Lampreys and amphibians have only superficial lateral line organs that do not form a deep lateral line system found in gnathostomes.

Lateral line hair cells show that each neuromast of a salamander is innervated by only two afferents and that each afferent contributes to only one of the two central fascicles present for each peripheral nerve (Fig. 16). Each neuromast of salamanders has hair cells of opposing polarity. Electrophysiological work has suggested that each group of hair cells polarized in a given direction is innervated by a single afferent. It has been hypothesized that each hair cell of a given polarity (Fig. 5) within a given line is presented in a single fascicle. It seems possible that the neurons giving rise to the two afferents, and possibly also the two opposingly polarized hair cell populations, are separated by different birthdates, as shown in teleosts. Further work has elaborated on details of lateral line directional innervation is working out in the zebrafish and has been demonstrated for utricle and saccule polarity innervation.

Opposing planar polarity of sensory hair cells and their selective innervation by afferent nerves is determined asymmetric fates by the combined action of a transcription factor, *Emx2*. *Emx2* is expressed in the rostrally positioned sibling, which has its kinocilium in a caudal direction. *Notch* ligands expressed in the rostral cell activate *Notch* in the caudal sibling, where higher *Notch* activation inhibits the expression of *Emx2* in the rostral direction. Ectopic expression of *Emx2* makes the kinocilia of all sensory cells assume a caudal polarity. Consistent with the *Emx2*-expressing cell, in *Notch1* mutants, all the sensory hair cells express *Emx2* and develop kinocilia with caudal polarity. Broad activation of *Notch* has the opposite effect: it inhibits *Emx2* expression, and it forces all kinocilia to adopt a rostral polarity (Ji et al., 2022). A bi-stable situation may determine by *Notch*-mediated lateral inhibition, the presence of *Emx2* in the rostral sibling determines caudal localization of the kinocilium. *Emx2* and *Notch* may compete for influence on some hypothetical factor that determines caudal localization of the kinocilium on the apical surface: *Emx2* promotes its action, while *Notch* inhibits it. As a result, the kinocilium was on the caudal side of the apical surface in most hair cells which shows interactions in different hair cell patterns of various amphibians (Engelmann and Fritzsch, 2022).

In summary, craniates have distinct mechanosensory hair cells that depend on *Atoh1* and other factors that receive lateral line innervation by at least two afferent fibers per hair cell bundle and opposing orientation that depends on *Emx2*. Despite the overall similarity with respect to stereocilia with asymmetric kinocilia among craniates, the lateral line hair cells are unique between cyclostomes and gnathostomes with respect to MET organizations: single *Tmc* (hagfish and lamprey) or two *Tmc's* (gnathostomes).

2.3. Evolution of electroreception: An early formation combined with multiple losses

It took a long time to understand the mechanosensory lateral line system that has so many similarities with the inner ear, which was typically referred to as the 'sixth' sense. In contrast, strongly electric fishes and amphibians, some of which are capable of producing numbing and shock, have fascinated the world for the last 5000 years: they have a unique 'seventh' sense that humans cannot perceive and have difficulties in understanding this alien sense. Investment with powers beyond our own perceptions engages humans in a profound curiosity. This curiosity was shared throughout antiquity, such as by Aristotle, Lorenzini, Volta, and Galvani. The difference was too distant from our level of understanding to comprehend electroreceptors. Attempts to explain electric organ function led to the invention of batteries and thus to bioelectric batteries, which may soon drive electric devices designed to replace dysfunctional organs. The battery provided the first source of continuously available electricity. This provided sources for electricity used by engineers such as Faraday, Edison, and Tesla, who started the technical revolution we are still enjoying today. The template for Volta's battery was based on the electric stingray organ as described by Lorenzini. The battery was meant to prove that bioelectricity was the same as the electricity generated with Galvani's metal plates. After this discovery, many used batteries to elicit sensory stimulations and even make cadavers twitch, much like Galvani had done earlier with his frog leg experiment. By creating his battery, Volta demonstrated that electric fish contain a battery that can provide readily available electric energy (Jung et al., 2022).

It was over 200 years ago when the first insights of an electric sense were proposed: they postulated the existence of a sense that was radically different from another sense humans possess that the electric eel was able to detect whether an electric circuit was open or closed. It took another 100 years before several people 'invented' the electric sense. The discovery of the electric sense highlights one dilemma for researchers: should they trust observations when they indicate unexpected or even alien phenomena?

More than 200 years later, we witness a second revolution again driven by electric fish: we may soon use our genetic understanding of electrocytes to generate electric organs in humans. The artificial organs will function like organic batteries and drive implanted devices such as cochlear implants or pacemakers. Moreover, the strong bioelectric fields produced by electric eels and rays lent insights into how the nervous system worked. With the invention of sensitive amplifiers and easily available recording equipment, Hans Lissmann (Lissmann, 1958) discovered a whole new world of fishes capable of producing and perceiving weak electric signals (Bullock and Hopkins, 2005).

2.3.1. Electroreceptor sensory neurons depend on common placodal origin

In contrast to the nearly ubiquitous lateral line system in vertebrates (except amniotes), electroreceptors are present in only some ~5,000 craniates whereas most of ~65,000 species have lost electroreception: Electroreceptors are absent in hagfish, most derived teleost's, all anurans, some salamanders, certain caecilians, and are absent in all amniotes. Electroreception shares similar organizations with lateral line fibers and projects exclusively with aLL, but not pLL fibers in basic vertebrates. However, in certain bony fish (silurids, gymnotids, mormyrids), electroreceptive afferents show a divergent pattern of innervation that can become enlarged in certain teleosts. It is thought that the evolution of electroreception may have evolved in parallel among the few bony fishes that are distinct in many features separating from most vertebrates. Electroreceptor fibers (ELL) project selectively by a distinct branch of the aLL (or multiple electroreceptor branches) to reach the dorsal nucleus (DN), also known as the dorsal octavo-lateral nucleus (DON).

Electroreceptors are well known in lampreys, chondrichthyes, silurids, gymnotids, mormyrids, bichir, sturgeons, lungfish, *Latimeria*, many salamanders and caecilians (Fig. 17). All electroreceptor afferents project exclusively by branches of aLL without contributing to pLL fibers (Jung et al., 2022). Branches can diverge to innervate together with pLL fibers like in lampreys. The exception is among silurids and mormyrids that project with both aLL and pLL fibers and, in gymnotids, have a unique recurrent aLL projection that is unique for the branches. Branches of electroreceptor afferents extend from the lateral line and project centrally to the dorsal nucleus (DN). The distinct projections of electroreceptor afferents to dorsal nuclei are likely homologous and are supported by electrophysiology. Electroreceptor innervation is well described for lampreys. Notably are the enlarged central projections of electroreceptor afferents that project rostral and caudal to extend between r2-r6.

Electroreceptor afferent fibers run independently of lateral line fibers, where they branch to single electroreceptors or clusters of electroreceptors. These electroreceptors are innervated by at least one afferent fiber (Fig. 17). Electroreceptive afferent fibers have no efferent innervations, unlike the efferent to the inner ears. All aLL and electroreceptive nerve fibers are devoid of myelin in lampreys and hagfish. Most likely, hair cells are innervated by individual nerve terminals on round bars or ribbons around 0.4 μm in diameter. Early investigations on the placodal development of lampreys require follow up to define the time and process of nerve fibers peripheral innervations.

Electroprojection is well described for many gnathostomes, but the central projections of *Latimeria* and lungfish have not been shown. Given that basic holosteans lack electroreception, the outgroup of teleosts, the evidence suggests parallel innervation of silurids, gymnotids, and mormyrids. There are unique projections of electroreceptive fibers with aLL and pLL among silurids, complicated projections with aLL and pLL in mormyrids, and projections with a branch of a recurrent aLL in gymnotids. Neuroanatomical data shows the presence of unique dorsal fiber projections in *Latimeria* and lungfish but requires tracing of nerve fibers to show the formation of distinct projections. Detailed descriptions are presented for the electroreception of salamanders (Fig. 17) and caecilians. Individual fibers from single electroreception fibers were labeled to project to the dorsal nucleus, distinct from two opposing mechanosensory projections for a given neuromast to the intermediate nucleus (Jung et al., 2022).

In summary, the dorsal nucleus receives innervation from electroreceptors and is homologous among craniates that have lost it in various vertebrates. At the periphery, electroreceptive hair cells receive innervation from at least one afferent that

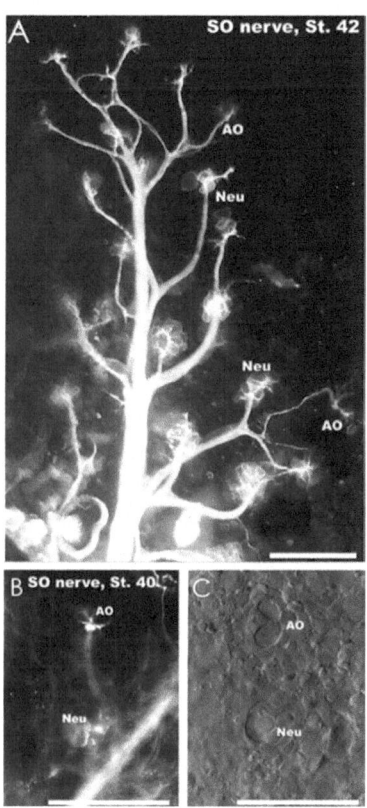

Figure 17: Peripheral fibers during development trace the lateral line and electroreceptors. Three branches have been identified to provide ganglia to the lateral line and electroreception: the ganglion of the anterodorsally, the anteroventral, and the dorsal branch (A). Detailed analysis of nerve fibers with lipophilic dyes shows the distinct innervation of an ampullary organ (AO, B, C) that can distinguish from the neuromasts (Neu, B, C). At a later stage, one can clearly identify the single fibers to ampullary organs (D) compared to two nerve bundles to the neuromast (D). Modified after Elliott and Fritzsch (2020).

lacks innervation by efferent. Electroreceptive nerve fibers are absent in hagfish, most bony fish, all amniotes, and all anurans. There is a novel gain in certain bony fish, which likely evolved in parallel at least twice (mormyrids, gymnotids, silurids). The extraordinary expansion of electroreceptors in mormyrids and gymnotids is extremely enlarged showing multiple unique features.

2.3.2. Electroreception of central projections innervate dorsal nuclei

Central projections are split into two groups. One group has high levels of similarity of electroreceptor projections to dorsal nuclei (lampreys, Chondrichthyes, bichir, sturgeon, *Latimeria*, lungfish, caecilians, and salamanders). The second group consists of silurids, gymnotids, and mormyrids that have unique branches of fibers and their central projections referred to as electroreceptor lateral line (ELL). Members of this group are so distinct that we touch on them only briefly as they have been presented exquisitely by others. Electroreceptors are absent in hagfish, most ancestral bony fish, anurans, and several other derived amphibians. Amniotes have lost electroreception, except for platypus in which electroreceptors are derived from trigeminal nerves.

The dorsal nucleus receives electroprojection in all major divisions of craniates, except for those lacking electroreceptors (Fig. 18). All electroreceptive fibers project through a single branch of aLL, which seems to be a common feature of craniates. All lateral line projections are entering near r4 and project rostral to r2 and caudal to r6. Among lampreys (Figs. 15, 19), the central projections show unique rostral and caudal expansions.

Chondrichthyes show distinct cerebellar-like expansion, the cerebellar crest, and show a dorsal granular ridge of the dorsal nucleus (Fig. 18). Among amphibians, only salamanders and many caecilians have a single projection together with the aLL to reach the dorsal nucleus (DN; Fig. 18). Rostro-caudal areas are for the dorsal nucleus are from r2 to r6, which is much shorter than lateral line fiber projections.

The three main central projections of silurids, gymnotids, and mormyrids highlight differences among them. **Silurids** have aLL and pLL afferents that provide electrosensory fibers. A cerebellar crista covers the DN and is known to project bilaterally to the midbrain (Fig. 18), the ventral torus nucleus. **Gymnotids** receive electric projections exclusively from aLL fibers but have a recurrent electric fiber system where they terminate in the electrosensory dorsal nucleus (Fig. 18). Electroreceptors use phase coders (T-units) and probability coders (P-units). Primary afferent fibers project T-units to spherical cells and project P-units to basilar pyramidal cells. Additional inhibitory granular interneurons add to the complex interaction. Different cell types lie between molecular and deeper electric fibers, comparable to cerebellar crest-like input. Four complete maps of the

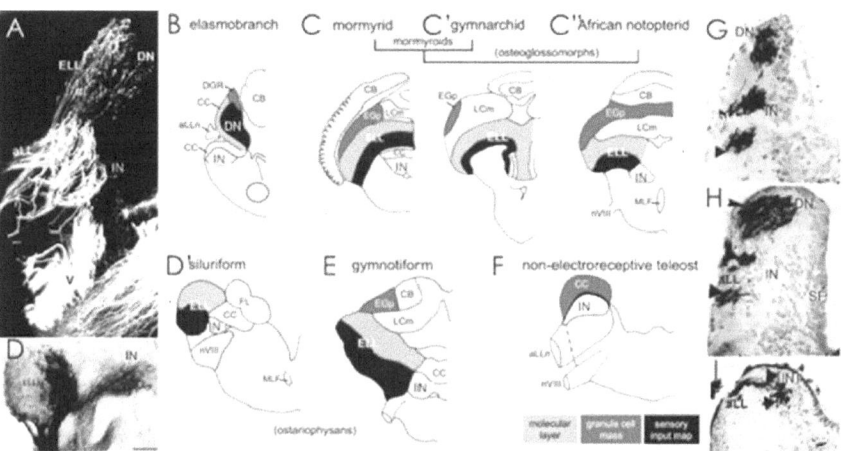

Figure 18: The lateral line in lamprey (A), various chondrichthyans (B), bony fish (C-F) salamander (G), caecilian (H), and Ascaphus (a frog, I). Black areas in D, G, H, and I indicate where primary afferent fibers terminate. Dark gray areas indicate granule cells from the molecular layer (light gray areas). aLLn, anterior lateral line nerve; CB, cerebellum; CC, cerebellar crest; DGR, dorsal granular ridge; DN, dorsal nucleus; EGp, eminentia granularis posterior; ELL, electrosensory lateral line lobe; IN, intermediate nucleus; LCm, molecular layer of the caudal lobe of the cerebellum; MLF, medial longitudinal fasciculus; nVIII, eighth cranial nerve. Fibers make distinct fascicles among some (A, G, H, I) whereas fibers may separate after the common fascicle projects (D). Note the absence of electroreceptors in most teleosts (F) and anurans (I). Modified after (Elliott and Fritzsch, 2020).

electrosensory surface are known with one projecting ampullary organs while the remaining three segments receive tuberous organs forming three maps on pyramidal and spherical cells. Pyramidal and spherical cells project to the lateral torus semicircularis and present different P-units and T-units. That information provides a complex interaction that may receive 6 neurons to provide information from and influence neurons dealing with P-type signals, indicating a first interaction of T- and P-systems. Diffuse inhibitory and topographic excitatory projections arise from the dorsal preeminential nucleus and terminate in the DN. Inhibitory bipolar cells terminate on pyramidal cell bodies, while excitatory stellate cells synapse in the lower molecular layer of the DN on pyramidal cell apical dendrites. **Mormyrids** have two electric projections: those of *Gymnarchus niloticus* and those of the remaining mormyrids that differ in electric projections with distinct knollenorgan and mormyromasts (Fig. 5). Electroreceptive fibers project along both aLL and pLL fibers to the equivalent of the DN. Unique to this group is that all mechanosensory lateral line fibers and electroreceptors project together to the DN, but with limited overlap in the DN and the cerebellar crista (eminentia granularis, (Jung et al., 2022)).

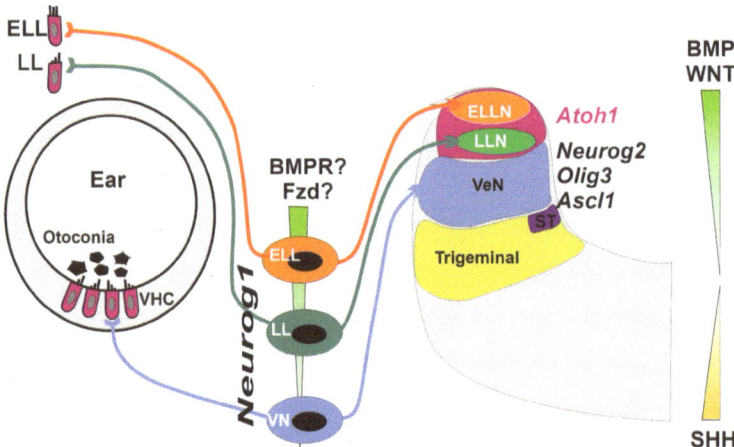

Lampreys, Chondrichthyes, Polypterus, Sturgeons, Latimeria, Lungfish, many Salamander, Caecilians, some Teleosts

Figure 19: Evolution from the common ancestor has led to the retention of the dorsal nucleus that depends on *Atoh1* to drive the central projection of electroreception. Different peripheral ganglia will depend on *Neurog1* and form a gradient of defining genes such as BMPs and Fzd for ganglia. BMP/Wnt (dorsal) and Shh (ventral) will interact with the brainstem. ELL, ELLN, electroreceptors and nuclei; LL, LLN, lateral line and nuclei; ST, solitary track; VeN, vestibular nuclei; VHZ, vestibular hair cells. Modified after (Elliott and Fritzsch, 2020).

Interestingly, both lateral line and electroreceptors project separately. The DN cortex forms three zones separate from ampullary organs and mormyromasts, each with branches in A-type and B-type fibers projecting to distinct areas and having a distinct projection of knollenorgans. The three DN inputs project to the optic tectum. Given that this entire lobe of electroreceptive and cerebellar nuclei is so transformed and enlarged that it is equivalent in size to mammals but is distinctive in its connections.

In lamprey and hagfish, some relevant gene expressions are known (Figs. 15, 18), including *Atoh1* and *Ptf1a*. As described by others in the development of mice, *Atoh1* is expressed dorsally along the spinal cord and hindbrain to the cerebellum in mice. Given the similar gene expression in auditory nuclei in mammals that replaces the equivalent nuclei of lampreys, we can assume that loss of *Atoh1* expression will eliminate electroreceptor and lateral line nuclei in lamprey (Fig. 19). However, lampreys and hagfish have a single *Lmx1a/b* gene, as compared a separate *Lmx1a* and *Lmx1b* in all gnathostomes. Following the data generated by *Lmx1a/b* double null mice suggests that neither lamprey nor hagfish would develop any electroreceptor or lateral line nuclei following the loss of the *Lmx1a/b* gene. There is an incomplete set of data for any additional *Wnt*s that may be needed in hagfish and lampreys.

Among bony fish, only zebrafish data have provided some clues about different genes in development. First, early gene duplication evolved into an *Atoh1a* and an *Atoh1b* that changes in early expression in the ear and brain. In addition, the ubiquitous early gene expression of *Sox2* necessary for early proneural bHLH gene expression shows interesting interactions of *Sox2* and *Sox3* and with downstream genes of *Atoh1*, *Neurod1*, and *Neurog1*. Interestingly, there is a unique projection of electroreception fibers that cross bilaterally at the dorsal fusion in mormyrids, somewhat comparable to the crossing fibers in *Lmx1a/b* double null mice. Whether *Lmx1a* and *Lmx1b* play a role in bony fish has not been analyzed.

Only recent data are provided for salamanders and frogs and show likely expression of *Atoh1*, *Sox2*, and certain other genes but central projections were not analyzed. Several *Lmx1a* and *Lmx1b* genes have been studied for early expression or very late in *Xenopus*. Different genes that could be eliminated and/or misexpressed are needed to fully understand the molecular basis for brainstem development of amphibians, including salamanders. The dorsal and intermediate nuclei overlap with *Atoh1* expression and likely depend on *Lmx1a/b* expression for their formation.

In summary, the DN projection is common among craniates. Central projections are complicated in certain cases with a cerebellar crest with and without segregation of distinct fibers. In contrast, all three electroreceptor teleosts are unique and provide the complexity of distinct multiple projections that are beyond the level of anything known among craniates (Figs. 19, 20). We are beginning a novel insight driven by newly uncharacterized genes that are essential for all dorsal parts of the hindbrain, including alar plate and choroid plexus formation. The common theme might be these genes are driving the development of the dorsal parts of the alar plate, including the dorsal nucleus, intermediate nucleus, and auditory nuclei.

2.3.3. Hair cell formation depends on *Atoh1* and other genes to develop electroreception

Electroreceptive organs were known for Chondrichthyes as the 'ampullae of Lorenzini' since 1678. These ampullary organs, found in various species, were proposed by the Sarasins to be 'electroreceptors' which was in contrast to the work of du Bois-Reymond that argued against this sensory system. Different features were found among different electroreceptor 'hair cells' and the earliest gain of new electroreceptors was described by Szabo, Bullock, Bell, Maler, and Hopkins, among others. Beyond earliest insights we have a well-established understanding of different electroreceptors, providing details on stereovilli (presence with few or numerous stereovilli or replaced with a single kinocilium) and showing unique innervation

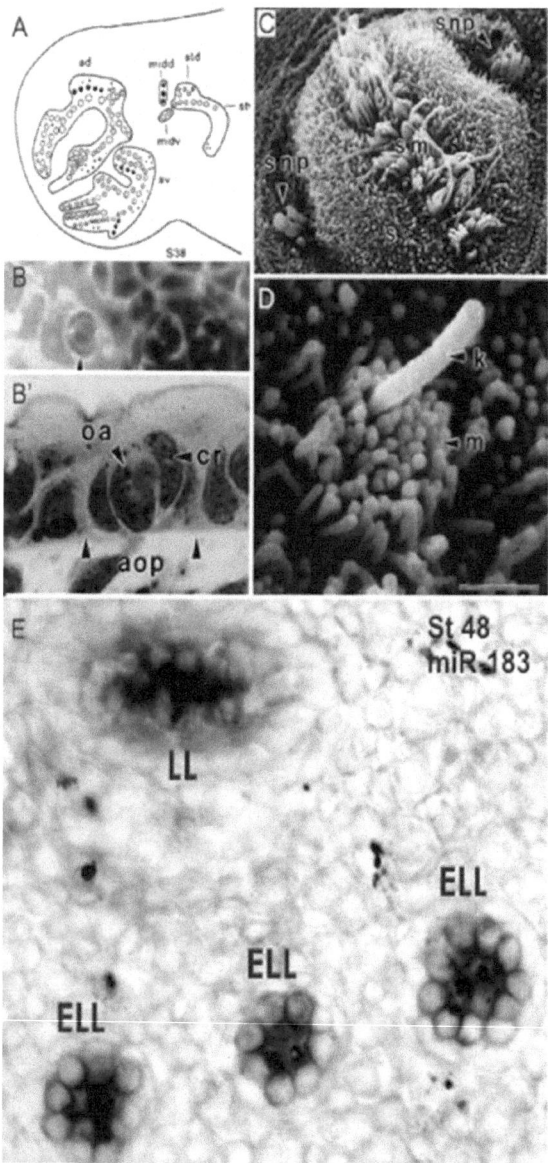

Figure 20: Details of lateral line (LL) and electroreceptors (ELL) in the axolotl. The first identification of electroreceptors (aop, ampullary organ primordium) has a single ampullary next to the lateral line (B, B'). Lateral line neuromasts are shown in an older animal compared to the earliest budding of ampullary organs adjacent to a kinocilium (E). miR-183 is expressed in early development of both lateral line (E, top) and electroreceptors (E, bottom). Modified after (Jung et al., 2022; Northcutt et al., 1995; Pierce et al., 2008).

by round or extremely elongated ribbons that are innervated by a single afferent fiber.

Interestingly, the various stereovilli and/or kinocilia indicate a large diversity of electroreceptors in bony fish. Electroreceptors of some species have a short kinocilium, with or without stereovilli (chondrichthyes, bichir, sturgeons, lungfish, caecilians). In contrast, adult lampreys and adult salamanders share multiple features like short stereovilli of many or few bundles and have no or very short kinocilia. Likewise, the three electroreceptive bony fishes have various stereocilia. In fact, the variety of multiple different organizations of all electroreceptive bony fish have shown a diversity of electroreceptors that is beyond the simple organization of lampreys and salamanders with respect to stereovilli.

A large diversity of electroreceptors is also indicated by the presence of various synaptic ribbons. Among the different synaptic ribbons in various electroreceptors are those that can be extremely large, round ribbons that are 2 μm in diameter, and those that can be small, round ribbons that are 0.2 μm in diameter. The ribbons are typically elongated and can be extensive and become very long.

Recently, the voltage-gated ion channel was identified for sharks, and it is distinct from gnathostomes. Silurids, gymnotids, and mormyrids have low resistance channels, whereas sharks, rays, bichir, sturgeons, salamanders, and caecilians have high resistance (Jung et al., 2022). Electroreceptor ion channels were identified as $Ca_v1.3$ in sharks and skates and may also be in salamanders. The multiple perspectives of electroreception have helped to understand its evolution. A different polarity is for non-teleosts versus teleosts: anodal positive to external in teleost versus cathodal positive internal for non-teleosts. Typically, electroreceptors are low frequency stimulation of 1-20 Hz whereas the tuberous teleost's are over 100 Hz or more frequency.

Electrophysiological evidence for electroreception was followed by neuronal tracing (Figs. 17, 20). Electroreceptors of lamprey have multiple stereovilli without a kinocilium and have a single round or plesiomorphic ribbons terminating on single, but branched, afferent fiber. The pattern of innervation and sensory organ identification of hair cells is not clear in lampreys. Only one type was identified as having large ribbons at the base with an enlarged single innervation showing 'multivillous' sensory hair cells in larvae and adults. Electroreceptors are likely independent 'multivillous' cells. From here one can conclude that electroreceptors function as independent receptors (lampreys), whereas they are aggregated into at least single sensory cells in others. Electroreceptors can occur in distinct areas that form larger branches in older larvae and adults that are all innervated by a single nerve fiber, with few cases of very large electroreceptors. Unfortunately, the multiple multivillous innervations by some were incomplete in the presentation from the original work. The multiple attempts to demonstrate the identification of 'hair cells' have not been successful (Jung et al., 2022).

Gnathostome electroreceptors have two basic patterns of hair cells: one with a short kinocilium surrounded by multiple taller stereocilia or with a single taller kinocilium surrounded by shorter stereocilia/microvilli. In sharks, rays, and ratfish, the Lorenzini's endings have a short kinocilium that extends from a basal body. In many ampullae, the kinocilium is surrounded by a few microvilli or can have short kinocilia with longer stereocilia. Synaptic innervation is from a single or a few afferent nerve endings.

Sensory cells are typically clustered in sensory ampullae and may have between 150-2000 different clusters. Bichirs are organized with a single kinocilium surrounded by short stereovilli that are particularly dense in the snout. Sturgeons form 5-30 branches of a single nerve fiber and have electroreceptors with a kinocilium and short stereovilli. Likewise, lungfish have a single kinocilium surrounded by a few microvilli. Electroreception of *Latimeria* requires a more detailed analysis of the unusual electroreceptors. Teleosts have a short kinocilium surrounded by multiple stereovilli. Various branches of the different hair cells of teleosts are a complicated shape among electroreceptors that have a diversity of stereovilli (Fig. 3) among all silurids, gymnotids, and mormyrids, and have a highly derived pattern of single afferents innervating oval or elongated ribbons. While electroreceptive bony fish respond to anodal stimulation, non-teleost's respond to cathode stimulation. *Xenomystus nigri* shows a single nerve fiber innervated by multiple elongated ribbons that are inside a deep depression. Caecilians have a single kinocilium with few microvilli. Among salamanders (Figs. 17, 20), different features have been identified, one with an apical kinocilia while others have multiple stereocilia. A notable difference is the enlarged ribbon of salamanders that can reach 3 μm in diameter compared to the very long ribbon of some teleosts.

An additional set of electroreceptors, referred to as tuberous organs, have a somewhat similar organization for silurids, gymnotids, and mormyrids: the sense organ is loosely packed on top of epidermal cells with protruding sensory cells open into an intraepidermal cavity. Most of the surface, the tuberous or knollenorgan is covered by large numbers of microvilli up to 3 μm in length and 0.1 to 0.15 μm in diameter. The sensory cell sits on a platform in which a single nerve fiber divides to form several boutons on the individual sensory cells, typically with shorter and more elongated ribbons.

Electroreceptive hair cells differ from those of the lateral line in terms of their requirement for innervation. Electroreceptors disappear rapidly upon cutting the nerve, followed by regeneration and normal development, whereas lateral line hair cells are not affected following nerve cutting. A similar dependency of innervation is shown for cochlear hair cells. Though cochlear hair cells disappear within several months, there is more variation in dependency for vestibular hair cells following innervation loss from neurotrophin elimination. Somewhat similar effects have been described in other bony fish, suggesting that electroreceptors and cochlear hair cells critically depend on innervation, whereas mechanosensory lateral line hair cells and vestibular hair cells respond to the loss of innervation with a very large delay. Interestingly enough, a similar rapid loss of innervation rapidly affects taste buds which we know depends on *Bdnf*. The bony fish have undergone early gene duplication, resulting in multiple neurotrophins. This makes it complicated to clarify which trophic support may be used in bony fish. Moreover, lampreys may have less complicated gene expression in *Bdnf* function and may also require other divergent neurotrophins.

Description of various early electroreceptors is incomplete in afferent neuron development as well as hair cell differentiation, with limited additional information on afferent innervation. Most likely, the electroreceptor hair cells will follow the general rule of mechanosensory line development that likely will depend on *Atoh1*, downstream of *Sox2*. The axolotl allows for the transplantation of tissue that can develop as both lateral line and ampullary organs. Likewise, transplantation of electroreceptors was demonstrated in sturgeons and sharks. These experiments demonstrated that the transplanted tissue could generate electroreceptors and lateral line neuromasts in non-sensory areas. How long of these transplantations persist and were reversed to innervate by lateral line fibers to innervate electroreceptors require work.

Recent studies of gene expression provided evidence for the evolution of electroreceptors as sister cell-types. Normal development of all lateral line, electrosensitive, and inner ear receptors requires micro RNAs. Loss of gene *Dicer* can affect not only the lateral line but also affects olfaction, the retina, and large areas of the telencephalon. Studying mechanosensory and electroreceptive system development could help to define the earliest expression of initial steps of mechano- and electroreceptor hair cells (Fig. 20).

In development, electroreceptors and lateral line hair cells can be close to each other (Baker, 2019; Engelmann and Fritzsch, 2022; Jung et al., 2022). Demonstrating adjacent distribution of lateral line and electroreceptors allowed us to identify the different organs, starting as singular electroreceptor cells (Fig. 17). Further work shows that each neuromast of a salamander is innervated by only two afferents and that each afferent contributes to only one of the two central fascicles present for each peripheral nerve.

In summary, electroreception is present in only some craniates, making a distinction within a given group of animals that have (lamprey, gnathostomes, amphibians) and do not have electroreceptors (hagfish, modern bony fish except silurids, gymnotids, and mormyrids, anurans and some derived salamanders and caecilians, and amniotes). Lamprey, silurids, gymnotids, mormyrids and salamanders share a similarity of electroreceptors with stereocilia as compared to using a kinocilium in electroreceptors of all other gnathostomes. Likewise, the plesiomorphic ribbon synapse is shared by lampreys, Chondrichthyes, lungfish, caecilians, and salamanders as compared to various elongated ribbons that have been found among derived bony fish. Electroreceptors are forming with short kinocilia and longer stereovilli or longer kinocilia and shorter stereovilli. In addition, they have a single afferent innervating the round or elongated ribbon synapses. Stereovilli are different between lampreys, derived bony fish, and most salamanders, which seem to be all similar in their overall kinocilia length, and sharks, rays, bichir, sturgeon, lungfish, and caecilians with short kinocilium and may lack stereovilli.

CHAPTER

3

Connecting a Novel Sensory Input from the Ear to the Brain: How to Make New Neuronal Networks and New Connections of an Auditory System

In contrast to the long history of the vestibular, lateral line, and electroreception (Chapter 2), the evolutionary origin of the **Central Auditory System** is a new formation in sarcopterygians:

1. A basilar papilla with a tectorial membrane that is in contact with the cochlear aqueduct is unique among sarcopterygians. A basilar papilla is lost in several lines (lungfish, many salamanders, some caecilians, and possibly one frog) and will develop in mammals into the organ of Corti.
2. Connections will receive and project from neurons, like vestibular neurons initially, that will eventually develop into unique fibers in frogs and sauropsids whereas it reaches out to spiral ganglia neurons of the cochlea in mammals.
3. Central projections of neurons reach out auditory nuclei only in tetrapods that develop into a tonotopic central projection in frogs and amniotes.

A novel system, hearing perception, is a unique sense of gnathostomes that shows an independent transformation from derived bony fish, the sarcopterygian, and the actinopterygians.

The auditory system relies on sound pressure that provides unique auditory inputs. The origin of sound pressure requires the lungs and associated choana which is the basis for air pressure. A parallel transformation of sound pressure is depending on bony fish that uses

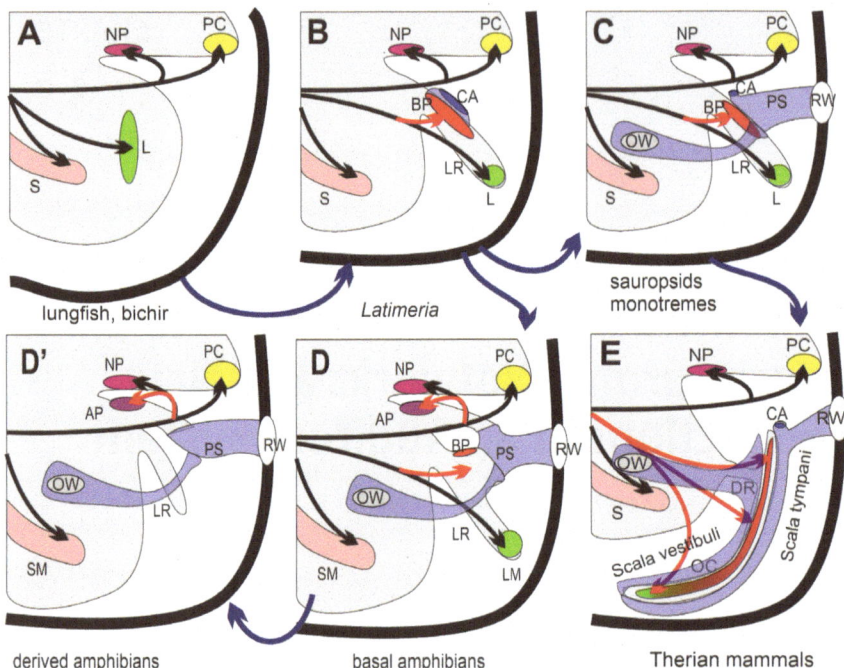

Figure 21: The evolution of the basilar papilla evolved from *Latimeria* and tetrapods. A lagena of lungfish and bichir show a common recess shared with the saccule and lagena (A). A separate recess (LR) forms the lagena and has a separate branch of innervation to reach the basilar papilla (BP) in *Latimeria* (B) and most tetrapods (C, D, E). Connecting with the cochlear aqueduct leads to an opening in *Latimeria* and amniotes (B, C, E). Therian mammals (E) may present a transformation of the lagena as integrated into the cochlea. An additional formation develops from amphibians, the amphibian papilla (AP; D, D') that splits from the neglected papilla (NP). Derived amphibians show a loss of BP and lagena due to a loss of lungs. Note that the connection with a recess in AP and BP in amniotes form a complex perilymphatic space. AP, amphibian papilla; CA, cochlear aqueduct; BP, basilar papilla; L, lagena; LR lagena recess; NP, neglected papilla; OW, oval window; PC, posterior crista; PS, perilymphatic space; RW, round window; S, saccule. Modified after (Fritzsch et al., 2023b).

different sensory epithelia without involving the sarcopterygian basilar papilla (Schulz-Mirbach et al., 2019).

I will provide an overview of the earliest formation of the auditory sensory organs, specifically the auditory neurons and the connected sensory epithelia (basilar papilla and amphibian papilla) followed by the formation of a unique central auditory system that evolved from ancestral vertebrates, after the loss of lateral line and electroreception (Fig. 9). Obviously, there is a formation of the lateral line and electroreception in most vertebrates (Section 2.2, 2.3), whereas amniotes lose the two sensory neurons, brainstem, and hair cells of these two sensory systems. Instead, a novel formation evolves an auditory system in most tetrapods. I will describe the transformation of the electroreceptor and lateral line central projection that will be transformed into the auditory system (Figs. 9, 21) and explain how the evolution of an auditory system can be explained by using examples for mammalian auditory development of mice.

3.1. The lungs evolved among sarcopterygians

Lungs will change in volume proportional to the far field oscillations. Lungs, trachea, larynx, mouth, and the presence or absence of a spiracular/tympanic membrane depend on the coupling of pressure to the ear that affects hearing. A novel hearing system and its function will depend on the basilar papilla that is common among sarcopterygians and tetrapods and shows a reduction in lungfish and several lines of amphibians. In contrast, a unique connection with the ear exists in derived teleosts. The pressure of the gas bladder (aka swim bladder) in bony fishes clearly affects the air directly or indirectly. Obviously, derived bony fish have evolved independently a hearing system that uses the swim bladder to provide for the hearing system (Schulz-Mirbach et al., 2019). I will first describe the evolution of the pressure of air first followed by the central and peripheral neuronal adaptations of the ear, the neurons, and the central auditory system, in tetrapods.

Without a doubt, a major novel event led to the lungs evolving early in vertebrates. The first vertebrate probably arose about 540 million years ago and may have evolved lungs at least 438 million years ago. Elasmobranchs have no lungs whereas sarcopterygians (Fig. 22) have bipartite lungs that start from a ventral tube of the esophagus. A recent set of data has identified close to 200 sarcopterygian fossils of which only four are recent species may have lungs: *Latimeria, Neoceratodus, Lepidosiren,* and *Protopterus*. I split the formation of lungs described in various fossil species compared to limited data on extended species that lately has been shown to have a small open lung in *Latimeria*.

Evolution is known in adult bony fishes but is undescribed in the earliest stages of development of sarcopterygians that emerges as a ventral protruding of trachea and lungs (Fig. 22). The anterior endoderm is supplied by the pulmonary artery and produces early surfactant proteins. We know that early development involves the formation of surfactant that allows the opening of the lungs to allow air for inhalation. Without inhaling the air, which is a first step for all sarcopterygians, including man, will not progress to develop a lung. Breathing the air first is a crucial step for almost all sarcopterygians except for secondary losses of lungs

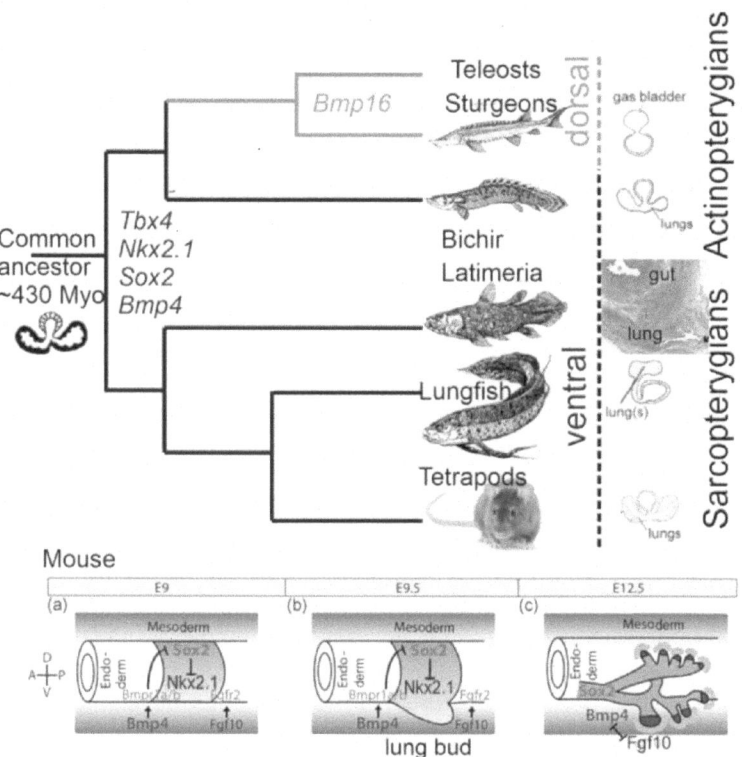

Figure 22: Lungs evolved early about 430 Myo ago. Evolution was an early formation in sarcopterygians and actinopterygians. The shared distribution forms ventral in early formation that depends on *Nkx2.1*, *Sox2*, *Bmp4* and *Tbx4* in sarcopterygian, and bichirs. Recent evidence suggests the presence of a ventral lung formation in *Latimeria*. A switch on the expression of *Bmp16* correlates with the formation of a gas bladder that comes from the dorsal part. We depicted the mouse gene expression that shows the initial upregulation of *Nkx2.1* that leads to the protrusion of the lung bud formation ventrally (bottom). Shaded gray indicates a gradient of *Sox2*, *Nkx2.1*, and *Fgf10*. Later stages drive the alveolars under the guidance of *Fgf10*. Images compiled from (Fritzsch et al., 2023b).

in amphibians. Evidence suggests that the earliest formation depends on a gene, *Tbx4*, that is clearly positive and expressed as the lung expands from the esophagus. Following the earliest formation of sarcopterygians and bichirs, recent evidence suggests that ventral budding is required. Based on the earliest expression of *Nkx2.1* and *Sox2*, both transcription factors, which is conveyed ventral budding (Fig. 22). Loss of *Nkx2.1* results in a tracheoesophageal fistula formation. In addition, *Bmp4* is also expressed ventrally during lung development and was recently described in mice, chickens, frogs, and bichir. A gain of highly conserved genes (e.g., *Nkx2.1, Tbx4, Bmp4, Sox2*) is in contrast to the loss of fewer highly conserved genes in sarcopterygians (*And1-3*), lungfish (*And1/2*), and tetrapods (no *And*), respectively [Fig. 22; (Amemiya et al., 2013; Wang et al., 2021a)]. Incidentally, a parallel loss of fins-to-limb and an increase/ decrease of expression in lungs of *Slc34a2* and a novel expression of *Foxp1* in sarcopterygians: actinotrichia proteins (*And1/2* and *3*) are common in gnathostomes, actinotrichia are reduced in lungfish (*And1/2*) and lost in tetrapods [Fig. 22; (Nakamura et al., 2021; Wang et al., 2021a)]. In summary, molecular data suggest that the lung buds form ventrally in sarcopterygians, lungfish, and bichirs, whereas dorsal expression leads to a dorsal expansion of a swim bladder in derived actinopterygians.

Interestingly enough, for *Latimeria*, I suggest a similar expression of *Bmp4* that splits from actinopterygians. Instead of showing a consistent formation in bichirs and tetrapods, it changes the expression of *Nkx2.1* and *Sox2*. A unique expression of *Bmp16* should regulate the dorsal formation of gas bladders in actinopterygians. Branching of another gene, *Fgf10* (Fig. 22), plays an important role for branches of the lung expansion. In summary, molecular data suggest that lungs bud ventrally in sarcopterygians and bichirs whereas a dorsal expression switches to a dorsal expansion in derived actinopterygians.

Coelacanth are described in some specimens that suggest the presence of lungs in some fossils. The length of lungs can reach 40 cm in *Mawsonia* specimen that can reach between 3-5 m in total size, significantly larger compared to about 1.5-1.8 m in *Latimeria*. Some fossil coelacanths show calcified lungs with one or two constrictions, interpreted as multichambered lungs.

Four areas were identified in *Latimeria*:

1. A zone of a pneumatic duct adjacent to the esophageal wall.
2. A diverticulum just outside of the esophagus.
3. An anterior chamber that does not allow dissociation from the esophagus and lung.
4. A residual cord.

The lungs are pseudostratified and pleated epithelium with columnar ciliated cells intercalated by goblet cells. In contrast to the arrested adult *Latimeria*, an early embryo has lungs that extend throughout the length. Extant air-breathing actinopterygians and sarcopterygian fishes are functional respiratory lungs. A single chamber is known in certain lungfish, amphibians, and the majority of adult Lepidosauria whereas in most amniotes a bipartite lung of various sizes forms that branch with numerous alveoli. A complete loss of several amphibians that have instead depended on vascular blood supply from the skin (Wake and Hanken, 1996; Wake and Donnelly, 2010).

In summary, the formation of lungs is primitive for sarcopterygians, including *Latimeria*. Consistency with the early formation from the esophagus to protrude ventrally depends on specific genes. In contrast, gas bladders are a dorsal formation in teleosts. The loss of all lungs is a secondary loss of caecilians, one frog, and multiple salamanders, the plethodontids.

3.2. Formation of a spiracle *versus* a tympanic membrane

A major distinction of the Eustachian tube (aka pharyngotympanic tube) is in the tympanic membrane (Fig. 23) that has a different function of the spiracle duct in many basic vertebrates or its absence in derived lungfish, and derived amphibians.

The function of the Eustachian tube is to protect, aerate, and drain the middle ear and permit the gas pressure in the middle ear cavity to open into the choanae to adjust external air pressure. Simply speaking,

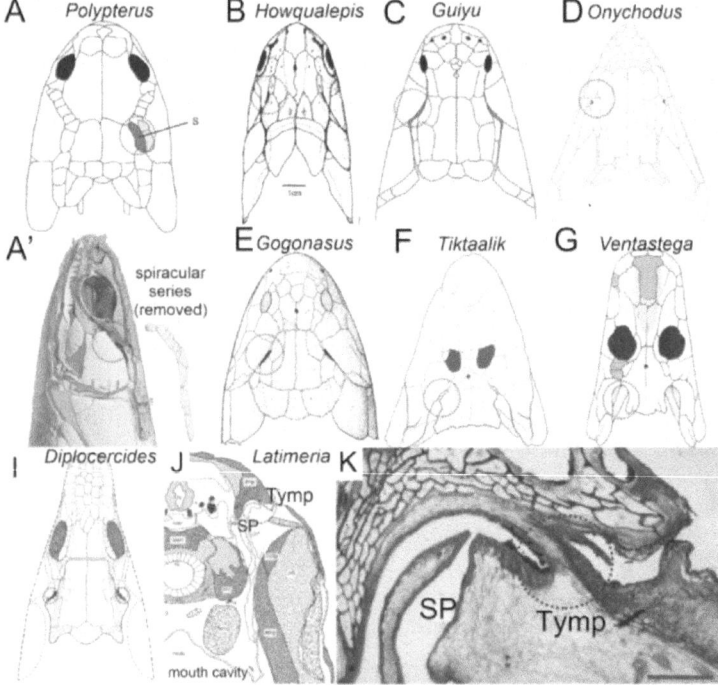

Figure 23: An opening of the spiracle (S) is found in lampreys, bichirs, and many fossils of Osteichthyes and fossil tetrapods (A-G). In contrast, a tympanic membrane (Tymp) is closed in *Latimeria* (H, I), showing the spiracular pouch (SP) extending from the mouth cavity up to the tympanic membrane. Modified after (Fritzsch et al., 2023b).

the Eustachian tube allows for the regulation of the pressure across the tympanic membrane that opens near the nasopharynx in mammals. One can show the free exchanges between the mouth (buccal cavity) that opens to the choanae freely to allow the larynx to connect the lungs to reach the tympanic membrane (Fig. 23). Choanae allow the larynx to influence the air connection between the lungs and the tympanic membrane, as demonstrated in frogs (Ehret et al., 1990).

A tympanic membrane is described that is found adjacent to the spiracular duct in *Latimeria* (Fig. 23), which is open to the adjacent space which extends to the mouth filled with water (Bernstein, 2003; Bjerring, 1985; Fritzsch, 1992). In addition, a complex interaction exists between the tympanic/stapes that are attached to the oval window on one hand and the perilymphatic opening near the round window on the other hand. The absence of a tympanic membrane is adjacent to *Latimeria* (Fig. 23). The transformation of the stapes that are associated with the oval window of amphibians and amniotes allows a push-pull situation between the perilymph opening of the round window and the tympanic membrane (Fig. 24) of the oval window. Obviously, a delayed development of stapes let in tadpoles to a direct sound driving the lungs that eventually evolve into the three ossicles in mammals.

Hearing in frogs is suggested to be driven by the round window directly from the lungs (Witschi, 1949), and many amphibians have an additional connection between the inner ear and the forelimb, except for caecilians (Fritzsch and Wake, 1988; Jaslow et al., 1988; Lombard and Hetherington, 1993). A round window and the cochlear aqueduct are present among amniotes (Atturo et al., 2018; de Burlet, 1934; Zeyl et al., 2022) whereas the equivalent to the cochlear aqueduct is near the basilar papilla in *Latimeria* (Fritzsch, 1992) and amniotes (de Burlet, 1934). It remains unclear how much fluid will be moved that could possibly drive the ear by pressure across the cartilage part of the ear (Bernstein, 2003; Bjerring, 1985; Fritzsch, 1992; Millot and Anthony, 1965; Rosowski, 2013).

In contrast, the spiracular duct is open in lampreys, elasmobranchs, bichir, and multiple fossils including several fossil tetrapods (Fig. 23) that can exchange either air or water across the spiracular duct. Flow of air or water can exchange without very little effect of pressure changes that do not result in driving the nearby ears by flowing through the nearby spiracular duct. The best description is provided by the bichir which demonstrates the flow of water or air through the spiracular opening. All derived teleosts have lost the spiracular pouch showing that the loss of this system is related to duplications of genes in modern teleosts.

A unique evolution that lost the spiracular pouch is found in derived lungfish (Fig. 24). A transient formation has been described in an embryonic lungfish that will disappear later. Lungs are known in adult lungfish but have lost any connection to the spiracular pouch that does

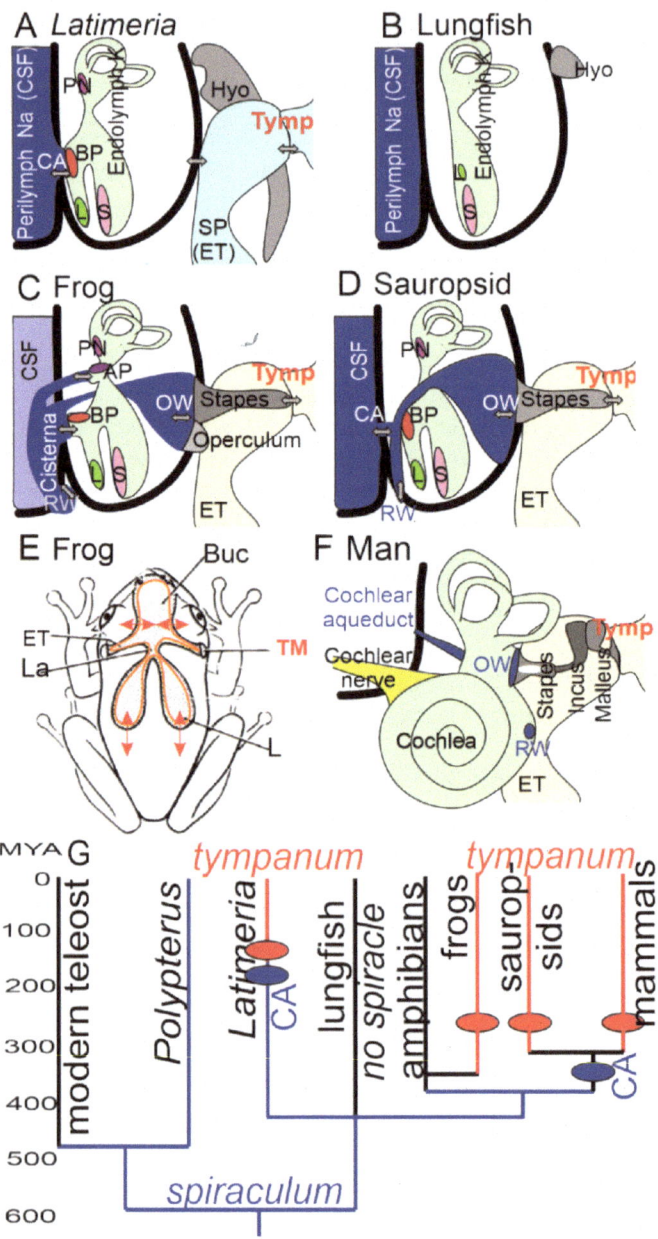

Figure 24: Comparison of the ear and tympanum, including a cladogram. The Eustachian tube (ET) leads to the tympanic membrane (Tymp) in *Latimeria* (A), frogs (C, E), and amniotes (D, F) but is absent in lungfish (B). Note that the fluid will be water (light blue; indicated by spiracular pouch; SP) while the tympanic membrane is surrounded by air (light yellow). In *Latimeria*, a cochlea aqueduct is directly connected with the BP. The hyomandibular bone is attached to the ear (A)
(*Contd.*)

whereas it is fused in lungfish (B). Lungfish have no basilar papilla (BP) but have a lagena (L) and a saccule (S). Various amphibians and amniotes have stapes that are inserted into the oval window (OW; C, D, F). Frogs and salamanders have a unique operculum. A push-pull dynamic driven by sound between the stapes, and the perilymphatic space (blue) is known in tetrapods, now indicated in the grey arrows also in *Latimeria*. Note a different access to the perilymph is indicated between frog (C) and sauropsid (D) that exits with a round window (RW). The cochlear aqueduct (CA) is known in amniotes but is absent in frogs. Air compression (E, inserted with red line) affects the perilymphatic space in amphibians (C, E) and amniotes (D, F). In humans, sound enters from the outer through middle and inner ear to reach the oval window across the three ossicles that exit into the Eustachian tube from the round window (F). In addition, the cochlear aqueduct makes it a 3-way connection between oval, round, and aqueduct. Bottom: A spiraculum evolved first (blue); it is eventually close to becoming a tympanum (red) in at least four sarcopterygian lines: Latimeria, frogs, sauropsids, and mammals. AP, amphibian papilla; BP, basilar papilla; Buc, buccal cavity; CA, cochlear aqueduct; CSF, cerebrospinal fluid; ET, Eustachian tube; Hyo, hyomandibular bone; La, larynx; L, lungs; OW, oval window; PN, papilla neglecta; RW, round window; S, saccule; TM, tympanic membrane. Modified after (Fritzsch et al., 2023b).

not affect the ear. Therefore, air does not connect with the ears in lungfish that have at the most a very small 200 Hz impact of sound. We know that several amphibians such as derived salamanders, caecilians, and one frog have lost lungs, a choana, and a spiracular opening. Obviously, high-frequency hearing (e.g., 2000 Hz) in frogs with a basilar papilla is absent in derived amphibians (many caecilians, many salamanders including the plethodonts). The amphibian papilla works at a lower frequency for hearing in frogs (Narins et al., 2006). A slightly higher frequency of 200-350 Hz is found in salamanders (Capshaw et al., 2020; Christensen et al., 2015b; Manteuffel and Naujoks-Manteuffel, 1990). Overall, except for frogs, most amphibians have a limited sound input, in particular the lungfish.

In summary, there is a major distinction between the spiracular duct in chondrichthyans, most actinopterygians, and many fossil sarcopterygians. The spiracular duct is lost in lungfish. In addition, several lines of amphibians have lost both the tympanum and the lungs. In *Latimeria* and tetrapods, the tympanic membrane is formed into four groups: *Latimeria*, frogs, sauropsids, and mammals (Carr, 2020; Fritzsch, 2003; Walton et al., 2017). Once sealed, the air or water will eventually generate a push-pull dynamic in the perilymphatic space of the ear.

3.3. The hyomandibular bone will be transformed into stapes

The hyomandibular bone was found in ancestral acanthodians some 438 Myo ago. It started as an addition to the spiracular foramen to form the hyoid bone. Various connections have been described as autostylic, amphystylic, and hyostylic connection that allows various ties between the hyomandibular bone and the skull. Used by many anamniotes, it functioned as an additional support for the jaws. Specifically, it appears that the hyomandibular bone/stapes formed after a gene expression is now clearer to develop in teleosts and mammals, respectively (Fig. 25). Evolution of agnathans and Chondrichthyes have diverged in connections with the skull. Expression of *Pou3f3* interacts with *Dlx1/2* to drive positively the hyomandibular bone/stapes whereas the *Dlx5/6* counteract the

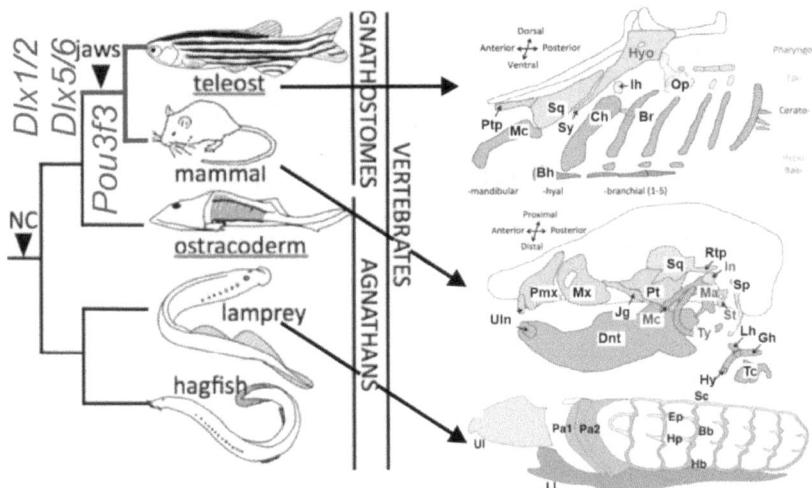

Figure 25: The relationships between jawed (gnathostome; teleost, mouse) and jawless (agnathan; lamprey, hagfish) vertebrates and urochordates. The last common ancestor of lamprey and hagfish diverged from the jawless fossil ostracoderms that likely split from specific genes for neural crest, in particular, *Pou3f3* and two closely associated genes, *Dlx1/2* and *Dlx5/6*: without these genes, the hyomandibular bone/stapes will not form. Skeletogenic neural crests build the mandibular arch and develop into the lower jaw and middle ear ossicles (malleus and incus) in mammals. The hyoid arch generates the jaw support and opercular (gill covering) skeleton in fishes and the stapes, styloid process, and hyoid bone in mice. Abbreviations: Zebrafish: Bh, basihyal; Br, branchiostegal ray bone; Ch, ceratohyal; Hyom, hyomandibular; Ih, interhyal; Mc, Meckel's cartilage; Op, opercular bone; Pq, palatoquadrate; Ptp, pterygoid process; Sy, symplectic. Mouse: Dnt, dentary; Gh, greater horn of the hyoid bone; Hy, hyoid bone; In, incus; Jg, jugal; Lh, lesser horn of the hyoid bone; lIn, lower incisor; Ma, malleus; Mc, Meckel's cartilage; Mx, maxilla; Pmx, premaxilla; Pt, palatine; Rtp, retrotympanic process; Sq, squamosal; Sp, styloid process; St, stapes; Tc, tracheal cartilage; UIn, upper incisor. Lamprey: Bb, branchial bars; Ep, epitrematic; Hb, hypobranchial; Hp, hypotrematic; Ll, lower lip; Pa1, first arch; Pa2, second arch; Sc, horizontal subchordal; Ul, upper lip; Vm, ventral pharynx. Modified after (Fritzsch et al., 2023b).

expression of *Pou3f3*. Evidence suggest that *Pou3f3* is initially expressed in the brain in lampreys but shows additional expression in the mesenchyme to initiate the formation of the cartilaginous formation of hyoid bones that will eventually expand to include the hyomandibular bone/stapes. Most important is a transformation of the hyomandibular bone into the stapes that has been co-opted to serve a radically different function as a middle ear ossicle in tetrapods.

Four transformations of the hyomandibular bone are known among gnathostomes:

1. The hyomandibular bone can function as a support for the first gill arch in the ratfish and elasmobranchs as a part of the hyoid bone to support gills.
2. *Latimeria* and other coelacanths have a long hyomandibular bone, comparable to basal Osteichthyes, and have an intracranial joint.
3. In modern lungfish the hyomandibular bone is reduced to a small appendix of unknown function; it evolved into a short hyomandibular bone that will be fused with the skull after they have lost the intracranial joint.
4. Tetrapods differ from non-tetrapods by transforming the hyomandibular bone by inserting it into the oval window, which allows directly driving the tympanic membrane with sound pressure. Obviously, the stapes evolved as separating it from its original bone function: instead of bracing the lower jaw it transmits vibrations of the tympanic membrane to the inner ear. A unique further transformation is in mammals that added two more bones to allow three ossicles driven (malleus, incus, stapes) by the tympanic membrane.

In amphibians, we can find three different levels of stapes formation (Jaslow et al., 1988). Starting in an amphibian organization, caecilians have stapes but have no operculum that connects with the limbs. In contrast, salamanders lack a tympanum, have stapes which in many salamanders can be fused to the ear (Szostakiwskyj and Anderson, 2022), and have the operculum to drive lower frequencies. Frogs are unique in terms of the delayed development of the stapes after metamorphosis that interact with the operculum. It is noticeable that the basal frogs have no stapes and will function like tadpoles using a lung to drive the sound pressure into the basilar papilla and the amphibian papilla.

Transforming the stapes from the hyomandibular bones must have happened as an insertion into the capsule of the ear by embryological regrouping in early development. *Dlx1/2, Dlx5/6* and *Pou3f3*: without all *Dlx*, and *Pou3f3* it will not develop a stapes or hyomandibular bone. How to insert the stapes into the otic capsule instead of abutting to the otic capsule is unclear and requires early association with certain early fossils.

In summary, three transformations of the ancestral hyomandibular bone can lead to the formation of the gill arch (ratfish), fuses with the skull (lungfish), and evolves into stapes (tetrapods). *Latimeria* has a hyomandibular bone that will function as in ancestral sarcopterygians, attached to, but not inserted into, the otic capsule.

3.4. Intracranial joint, basicranial muscle, notochord, ductus communicants, and basilar membrane are interacting

Undoubtedly, explaining the various possible interactions of the tympanum and ears does not fully explain the function of the role of lungs in sarcopterygians; a novel approach is needed. Here I try to connect various features into a single explanation.

The intracranial joint (IJ) allows pitching mobility between the otoccipital and ethmosphenoid portions. The IJ is found in all sarcopterygians except for lungfish and tetrapods, which have lost this joint movement. Neither chondrichthyans nor teleosts or lampreys have an IJ. The development is described in the earliest division of the IJ. There is a notable small brain volume that is filled with fluid, making it a unique size relative to the other sarcopterygians. How the evolution started in ancestral sarcopterygians about 415 Mya is unclear. The IJ allows for movements with a great bite for about 5 degrees down. Unfortunately, the degree of an upward movement is unclear as it does not happen by a dedicated muscle. Consequently, whether or not the gap can be increased or only decreased is not known. Future research is needed to fill in the gap that was suggested of up to 20 degrees. In summary, all sarcopterygians have a unique joint that develops early which allows great biting, but its early evolution remains to be seen (Fritzsch et al., 2023b).

The IJ allows primarily a strong bite down that uses the basicranial muscle (BM) to allow the contraction of the IJ. This unique muscle is innervated by the abducens that connects across the IJ. Like the IJ, the evolution of the BM is likely common among sarcopterygians. Tetrapods have evolved a muscle that is innervated by the abducens, the retractor bulbi, which connects to the eye for retraction. Retractor bulbi is unknown in lungfish, teleosts, and Chondrichthyans (von Bartheld, 1992). Lampreys have a different configuration with four eye muscles to move the eye like in gnathostomes. However, two eye muscles receive a distinct innervation by the abducens whereas a single eye muscle is innervated by the abducens in gnathostomes. It is possible that splitting the ocular muscles from three (lampreys) to four (gnathostomes) muscles (CNIII), could establish the gnathostomes organization, allowing for one trochlear (CNIV) and one abducens (CNVI) innervation across vertebrates. In summary, the BM of sarcopterygians evolved likely from an abducens innervated eye muscle in lampreys to eventually transform into the BM in sarcopterygians and become the retractor bulbi muscle in tetrapods.

The notochord is unique among sarcopterygians in that it is larger compared to that of other gnathostomes (Dutel et al., 2019). It was proposed to allow a recoil after the BM activation of the IJ to expand following contraction and relaxation. Additional evidence supports the recoil effect after the IJ is compressed by the BM movements. Developing data suggest a position starting in the ethmosphenoid to ending up in the occipital part where it remains close to the ear and the abducens nerve innervation allowing the movement of the IJ between the two parts of the skull. A notable increase in the diameter of the notochord during development shows an expansion from about 1 mm in embryos to close to 30 mm in the adult. Moreover, latest endocasts show a very large space compared to the size of the brain. In summary, the unique size and position next to the IJ of the notochord could function with the BM that allows fluid around the endocast.

A tubular perilymphatic passage that communicates with the basilar membrane has been described in certain detail. Millot and Anthony (Millot and Anthony, 1965) defined an endolymphatic fluid that is continuous with the lagena. However, detailed analysis showed that the *Ductus communicantis* (referred to here as cochlear aqueducto) is closed and to attached with the basilar membrane in *Latimeria*. Instead, it allows for communication between the cochlear aqueduct and the inner ear that uses a specific mix of potassium for function. The duct forms within a transverse groove adjacent to the occipital region and the notochord extends below the medulla oblongata. It extends across the membranous labyrinth where it contacts the otic capsule, specifically it is in contact with the basilar membrane and extends out the IJ. A similar organization is found in *Nesides*, a Devonian coelacanth that has a much larger foramen magnum compared to *Latimeria* (Bjerring, 1985). Given that fluid flow between the IJ and the foramen magnum should allow for some pressure to affect the basilar membrane, this duct may serves as release from pressure via the cartilage and the tympanic membrane, comparable to the function in tadpoles driven from the round window across the cartilage. How much fluid could flow and in which direction it will move as well as how much pressure it can exert within the communication and across the ear remains unclear. In summary, a possible connection exists between the foramen magnum with the basilar membrane that allows a perilymphatic fluid to move between the adjacent lungs and tympanic membrane and extends next to the IJ.

Undoubtedly, the basilar membrane is likely a homolog in sarcopterygians that may have evolved as a pressure and low frequency detector. Connecting the IJ with the notochord, the size of the endocast, the foramen magnum, and the round window with the basilar papilla and the tympanic membrane remains to be seen.

3.5. The role of air and sound in water

A clear perception of sound affect the ear by the lungs which will obey the gas laws and change volume when pressure is changing (Fig. 26). Changes in volume will pressure waves, to radiate from them (Popper et al., 2021). Pressure-to-displacement will transform the lung/tympanic membranes and the perilymphatic system to increase the pressure waves to be driving the ear. Fishes have special structures that connect mechanically the gas

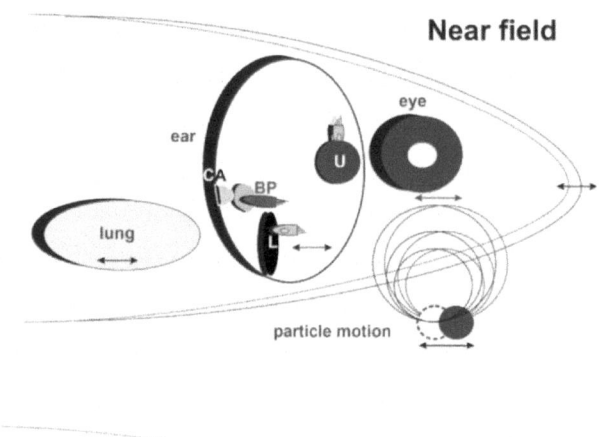

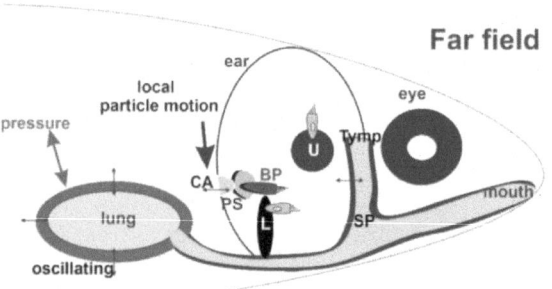

Figure 26: The direct particle motion will drive all sensory organs, including the eye and the lung, the same direction generating the near filed effect of a nearby stimulation. The fear filed is acting primarily in the lung and associated connections, including the spiracle, cochlear aqueduct (CA) and reaches out the perilymphatic space (PS) in *Latimeria*. Note that the pressure will oscillate in the local particle motion causing local effects. We assume that the pressure will reach the basilar papilla (BP) that would be driven in *Latimeria*, consistent with the fact that have a disruption of *Latimeria* consistent with a rapid pressure change. Modified after (Fritzsch et al., 2023b).

bladder to the ear to sound pressure, such as the Weberian ossicles, which affect a wider frequency range.

The impedance of water will depend on temperature, atmospheric pressure, solutes, and other factors. It is important to understand that a 36 dB reduction will happen to a completely exposed head, compared to the same pressure underwater, to affect the air: a tympanic membrane will reduce in the ear in water that will sound pressure in the air compared to water.

Obviously, the effect of air flowing through the spiracular pouch will have a limited effect on air passing near the ear. Once an air bubble can form between the choanae and the tympanic membrane, such as the tympanic membrane in *Latimeria* and most amniotes, will follow the pressure rules (Fig. 26).

A spheric air bubble (A) will affect pressure changes and will change its volume (ΔV), resulting in the displacement of air bubbles (d). Defining the quotient of volume and area will result in d=ΔV/A that will extenuate with an asymmetric air bubble and drives the perilymphatic system with or without a stapes attached to them. Note that the parallel formation of a gas bladder (=swim bladder) is a formation in derived bony fishes that follow the same rules of pressure-to-particle-motion transducers that radiate sound energy to the inner ears.

Latimeria has an open connection between the tympanic membrane and the esophagus [Fig. 26; (Bemis and Northcutt, 1991; Bernstein, 2003; Fritzsch, 1992)] that is comparable to the Eustachian tube of frogs, sauropsids, and mammals (Carr, 2020; Ehret et al., 1990; Grothe et al., 2004; Tucker, 2017). The tympanic membrane can provide oscillating pressure in *Latimeria*, supposedly without the use of the hyomandibula/stapes (van Bergeijk, 1967). In tetrapods, these oscillations could be driven by air pressure changes (Fig. 26) comparable to developing frogs via the round window (Witschi, 1949). A connection from the occipital region of the brain to the cochlear aqueduct exists in fossil sarcopterygians and in *Latimeria* (Bernstein, 2003; Bjerring, 1985; Fritzsch, 1992; Millot and Anthony, 1965) as well as in tetrapods (Atturo et al., 2018; Harper and Rougier, 2019; Wichova et al., 2019; Zeyl et al., 2022). All captured *Latimeria* were rapidly lifted from ≥100 m depth or deeper (Forey, 1998). Sudden air expansion in the lung and adjacent spiracular duct will disrupt the basilar membrane (Figs. 24, 26; (Fritzsch, 2003; Millot and Anthony, 1965)) that died shortly after bringing up the fishes (https://en.wikipedia.org/wiki/Latimeria#Biological_characteristics).

In summary, the choanae and the closed tympanic membrane of *Latimeria* may affect the ear across the cochlear aqueduct by sound pressure causing a push-pull dynamic interaction across the ear. Barotrauma in humans (Elliott and Smart, 2014; Hornibrook, 2018; Shupak, 2006) follows the same principle as observed in *Latimeria*, and disrupts the basilar membrane/cochlear aqueduct in *Latimeria* and man.

3.6. Evolving the ear shows loss and gain in sarcopterygians

Vestibular sensory organs are the most probable steps during the early stages of vertebrate auditory evolution. Lampreys and hagfish have a different distribution of sensory epithelia of ears compared to gnathostomes: cyclostomes have only two canal cristae and a single common gravistatic sensors (Fig. 14). Gnathostomes are characterized by three sensors for angular acceleration (the anterior, horizontal, and posterior (caudal) cristae) and has three gravistatic receptors (utricle, saccule, and lagena). I concentrate on sarcopterygians, amphibians, and amniotes, depicting the sensory evolution that evolved from the lagena. A lagena is absent in chimaeras, the ratfish, while elasmobranchs have a lagena that has a polarity compared to that of tetrapods (de Burlet, 1934; Lewis et al., 1985) is a part of the saccule. An incomplete split of lungfish and bichir is found in a lagena that is close to the saccule (Fig. 14). At least three independent segregations of the lagena forming its tube in derived elasmobranchs, derived teleosts, and derived sarcopterygians (*Latimeria* and tetrapods; Fig. 14). An amphibian papilla, used for sound perception in amphibians, can be considered as an exclusive transformation of the papilla neglecta (Fritzsch et al., 2013; Fritzsch and Wake, 1988; Lombard and Hetherington, 1993; Smotherman and Narins, 2004). In addition, the papilla neglecta in elasmobranchs can provide sound detection (Chapuis and Collin, 2022; Corwin, 1981).

Unfortunately, the resolution in fossils shows just the two or three semicircular canals, the utricle, saccule, and occasionally the lagena, but not any basilar papillae, nor the amphibian papilla in amphibians (Clack and Ahlberg, 2016; Zhu et al., 2021). Most recent data are expanded our fossil understanding of tetrapods (Harper and Rougier, 2019). Still, the obvious difference in the resolution of fossil ears, can be compared to much older ear images, including the pattern innervation, provided by Retzius (Retzius, 1881; Retzius, 1884). The details of the hair cells, tectorial membrane, and cochlear aqueduct and their innervation can be seen in *Latimeria* (Fritzsch, 1987) compared to the best images of whole skulls and fossil ears (Harper and Rougier, 2019; Zhu et al., 2021).

A unique feature forms in *Latimeria* that provide its own branch of the lagena innervation that sections in multiple fascicles, the basilar papilla (Fig. 27). It receives a direct connection to the perilymphatic space of the basilar membrane that is a common innervation of *Latimeria* and tetrapods (Fig. 27), innervating a separate branch to the lagena. The size

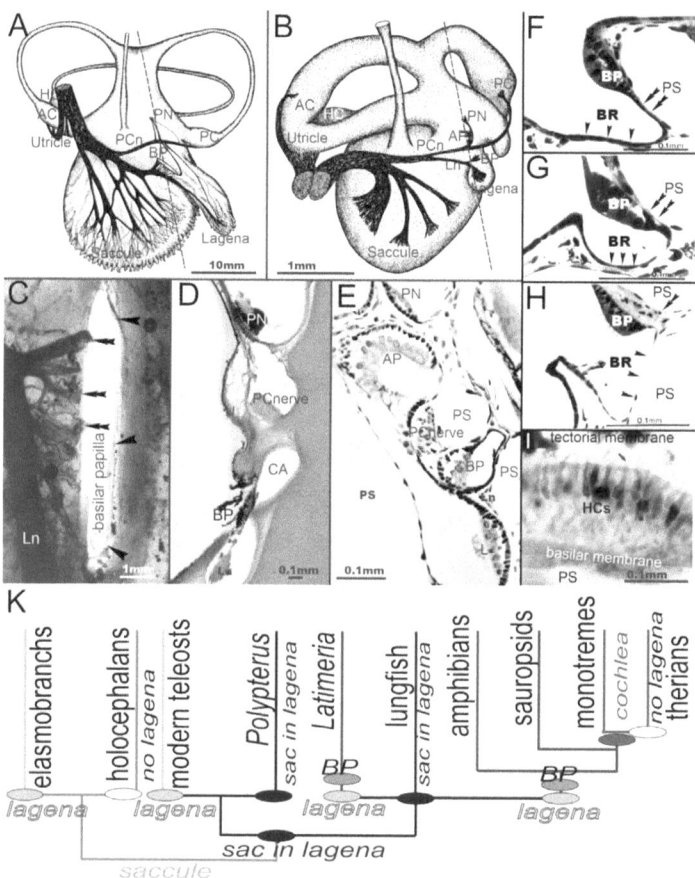

Figure 27: Latimeria resembles amphibians in the innervation pattern of the inner ear, including with cladistics. Whole image of the left ears comparing the lagena in *Latimeria* (A) with that of a caecilian (B). A separate nerve branch (Ln) innervates the basilar papilla (B, C, E). The space occupied by the basilar papilla is about 6 mm long and open to the cochlear aqueduct (CA; C, D, E). A papilla neglecta (PN) in amphibians transforms into the amphibian papilla. Note that the basilar recess (BR) is adjacent to the perilymphatic space (E-H) in amphibians. About 16 hair cells are shown in the cross-section of the basilar papilla of *Latimeria* (I); they are slightly larger than those in the amphibian papilla (K). Cladistic analysis suggests the earliest formation of a lagena in its recess versus the lagena developing in the saccule. A basilar papilla is correlated with the formation of the lagena in *Latimeria* and tetrapods. Note that the cochlea is transformed from the BP and loses the lagena in mammals and is shown as unresolved connections among sarcopterygians. AC, anterior crista; AP, amphibian papilla; BP, basilar papilla; L, lagena; Ln, lagena nerve; PC, posterior crista; PCn; posterior crista nerve. Dotted lines in (A, B) indicate the plane of sections of *Latimeria* and caecilian (D, E) compared. Modified after (Fritzsch et al., 2023b).

of the basilar papilla is about in 6 mm length and 1 mm wide, comparable to the size of many tetrapods including the size in crocodiles and small mammals (4-6 mm in length). I counted about 500 hair cells in *Latimeria* comparable to the hair cells in birds and mice (Manley, 2018). Loss of ear specific organization in lungfish should be considered a secondary loss of the tetrapod and *Latimeria*, highlighted by showing a common lagena/saccular recess that shows hardly any effect of sound in lungfish.

Unfortunately, in the absence of proper nerve fiber labeling with connections, as shown in *Latimeria* (Fritzsch, 2003), Bernstein (2003) indicated a new and wrong position of the basilar papilla, referred to as 'end organ'. The tectorial membrane is unique in most sarcopterygians, but the molecular analysis is partially clear (Mulry and Parham, 2020). A common feature in all sarcopterygians is the formation of a cochlear aqueduct adjacent to the basilar papilla (Fig. 27). The endolymph, which has a high level of potassium needed for hearing function, is separated from the perilymph that is continuous with the brain case and contains a high level of sodium produced in the cerebrospinal fluid [CSF (Szeto et al., 2022; Wangemann, 2002)]. In *Latimeria*, it was suggested that there is a communication of the endolymph of the ear with the adjacent perilymph (Millot and Anthony, 1965). The thin basilar membrane adjacent to the cochlear aqueduct in *Latimeria* is disrupted, likely due to barotrauma (Figs. 24, 26) which will allow the perilymph to mix with the endolymph. Likewise, barotrauma can exit from the cochlear aqueduct (Elliott and Smart, 2014; Wichova et al., 2019). Combined, the information suggests the basilar membrane is functional and responds to pressure changes (Bernstein, 2003; Fritzsch, 2003) and likely responds to lower frequency sound (Zeyl et al., 2020; Zeyl et al., 2022). The lack of ear-specific organization of the basilar papilla in lungfish should be considered a secondary loss (Fritzsch, 1992), highlighted by the common lagena/saccular recess in lungfish (Platt et al., 2004), which hardly show any perception of sound (Christensen et al., 2015a).

Caecilians, salamanders, and frogs belong to basic amphibians (Fig. 26). In contrast, derived amphibians show a reduction and loss of the basilar papilla in caecilians and salamanders. For example, no basilar papilla is absent in *Siren, Necturs, Notophthalmus*, and all Plethodonts. Amphibians can hear even without a basilar papilla (Capshaw et al., 2022). The basilar papilla can be short or long but follows the same basic organization that allows the exit from the three perilymphatic openings: the oval window, the round window, and the cochlear aqueduct (Manley and Sienknecht, 2013; Zeyl et al., 2020). A unique transformation is described in monotremes that show a separate formation of a lagena and a basilar papilla. In contrast, mammals have no distinct formation of a lagena that may become integrated into the apical part of the cochlea. We

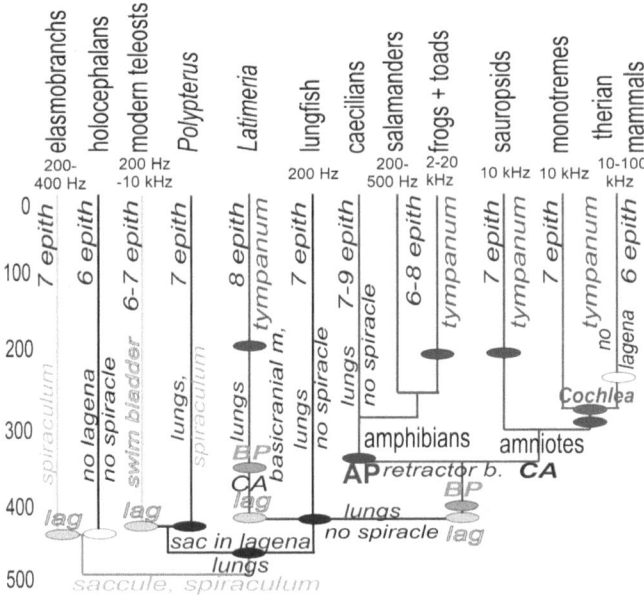

Figure 28: Hearing range and relevant features shown in a cladogram. Most gnathostomes have a low hearing range (200-500 Hz). Some elasmobranchs can hear up to 400 Hz and some teleosts reach above 10 kHz. Frogs can reach 20 kHZ, sauropsids and monotremes reach 10 kHz whereas certain mammals reach over 10 kHz. Note that with or without lungs the hearing range is nearly identical among gnathostomes except for adding the higher frequency in modern teleosts, frogs and toads, sauropsids, and mammals, with highest frequencies up to 100 kHz known for bats, whales, and clupeids. Papilla neglecta is incompletely listed. The number of sensory epithelia in the inner ear is highest in caecilians (9) whereas the inner ear of mammals and holocephalans have 6 epithelia. Note a trend to reduce the number of epithelia in derived amphibians. The formation of the tympanum occurred at least four times in the sarcopterygian lineage (*Latimeria*, frogs, sauropsids, mammals). In contrast, a spiraculum is found early in many recent and extinct gnathostomes. A basilar papilla adjacent to the round window (RW) is found in many recent and extinct sarcopterygians but has been lost in lungfish. Amphibians have a unique amphibian papilla whereas the cochlea evolved from a BP in mammals. A lagena forms in its recess (lilac) four at least three times independently (derived elasmobranchs, derived teleosts, sarcopterygians) whereas a lagena forms within the saccule. A lagena is absent in therian mammals and ratfish. Lungs appeared first in osteichthyans and evolved into the swim bladder in modern teleosts. A unique innervation of *Latimeria* is innervated with a basicranial muscle that seems to relate to the retractor bulbi of tetrapods; both muscles are innervated by the abducens and are absent in lungfish. Note we indicate the ages of gnathostomes and suggest the relationship between *Latimeria* and lungfish as unresolved. AP, amphibian ampulla; BP, basilar papilla; epith, epithelia; lag, lagena; retractor b, retractor bulbi; RW, round window; sac, saccule. Modified after (Fritzsch et al., 2023b).

assume that therian mammals are derived from development that shows a unique progression from the apex to the base.

A separate formation leads to the formation of the amphibian papilla (Figs. 14, 26). I suggest that the formation of the amphibian papilla is a unique transformation of the neglecta papilla. Note that the branch to the amphibian papilla shifts its innervation in derived anurans. The basilar papilla is reduced and is absent in derived caecilians and lungless salamanders that may also reduce the amphibian papilla and the lagena.

In summary, I describe the formation of the basilar papilla, tectorial membrane and the cochlear aqueduct that is found in *Latimeria* and most tetrapods that are topologically closely related to the lagena (Fig. 28). In contrast, lungfish and bichirs show a formation of the lagena and are incompletely segregated from the saccule. *Latimeria* has evolved a basilar papilla, comparable to tetrapods, who is lost in lungfish; the latter of which lack a basilar membrane, lack all hair cells, and all neurons targeting the basilar papilla. A basilar papilla is not only lost in lungfish but is also lost in derived amphibians. This contrasts with most tetrapods, which have a basilar papilla that is receiving a perilymph connection via the cochlear aqueduct, much like in *Latimeria*. It is possible that *Latimeria* can perceive sound as is known among all extant sarcopterygians that would be transformed over time into sound hearing. Essentially, the formation of *Latimeria* follows from the basilar papilla/cochlear aqueduct from inside-out to reach out via the lateral cartilage and the tympanic membrane. In contrast in amphibians and amniotes the sound pressure will hit in the tympanic membrane first, is driven via the stapes to hit the oval window where the sound will be converted into perilymph movement and will exit via the round window through which it will reach the basilar papilla/organ of Corti, an outside-in progression. Lungs to breathe air evolved early in osteichthyans to respond to sound and other pressure using the tympanic membrane (Fig. 28) in at least four sarcopterygians: *Latimeria*, frogs, sauropsids, and mammals. Depending on whether the spiracular duct is open or closed, air/water will be allowed through the lungs or may allow water through. In contrast, *Latimeria* and most tetrapods have a closed tympanic membrane that potentially could trap air or locally generated gas, affecting the nearby ear by pressure oscillations. Assuming the air bubble is working, the stapes will eventually be inserted into the ear to drive a push-pull dynamic from the perilymphatic spaces from the round window to the tympanic membrane in tetrapods. How the unique features of *Latimeria*, the intracranial joint, basicranial muscle, enlarged notochord and a very large foramen magenta, could allow the flow within the endocast, that is a small portion of the brain, remains unclear.

CHAPTER 4

Defining the Auditory System of Tetrapods

Owen coined the word **homology** to characterize identical organs, no matter what variation in form and function they may have in different species. Owen (Owen, 1848) wanted to distinguish those organs from organs that look alike because functional requirements have shaped them that way, identified as an **analogy**. The middle ear featured extensively in the early specification use of the term homology when Reichert (Reichert, 1837) showed that indeed two of the middle-ear ossicles were derived from jaw bones. The definition of **homology** has shifted from classical morphology to a more evolutionary definition that relies heavily on gene information flow, meaning genetic material that governs the development of a given organ. Transcription factors not only are homologous based on their sequence similarity, but also guide homologous developmental processes in otherwise dissimilar organs. If we accept such homology based on genetic identity of developmental modules, we will have to appreciate that apparently dissimilar organs within and across phyla might be derived from common primordia and thus can be homologous (Grothe et al., 2004).

Of course, we must keep in mind that, much like in the case of functional similarity (analogy or homoplasy), such developmental modules can be used in different organs and do not necessarily prove the homology of entire organs but only of genes and their developmental modules. For example, the organ of Corti is composed of three outer hair cells and one inner hair cell extending 6-24 mm long. In contrast, in monotremes, we can have up to 10 or more outer hair cells and up to four or five rows of inner hair cells that are much shorter and have a lagena near the apex (Schultz et al., 2017). Compared to a graduate change in the length of hair cells in birds and a different organization in the basilar papilla of tetrapods, we use mammals not as a basilar papilla due to the different organization of hair cells (Luo and Manley, 2020). In addition, most of the inner hair cells receive the type I spiral ganglion neurons whereas only ~5% of type II fibers reach out the outer hair cells, extending from the apex toward the base for several rows. Again, the pattern of innervation is different compared to most tetrapods making it the unique designation of spiral ganglion neurons in mammals but are all considered as the derived organ of Corti from the basilar papilla in sarcopterygians.

Other genes, which likely are not homologous across phyla, are needed in addition to forming unique organs. For example, the fibroblast growth factor *Fgf10* is essential for the formation of limbs and lungs and is involved in the formation of the semicircular canals of the ear (Fig. 16). Moreover, together with a second factor, *Fgf3* and *Fgf10*, is needed to help generate a placode and transform it into an otocyst in mammals. In addition, we know an early expression of *Foxi1* is needed for the earliest formation in collaboration of *Fgf3/10*, *Pax2/8*, *Sox2*, and *Eya1* to move forward. However, a distinct deletion of specific genes is needed for the cochlear ear, including *Gata3*, *Shh*, *Pax2*, and *Lmx1a/b*. In contrast to the interlink of cochlear hair cell formation, we can add to the function of *Neurog1*, *Sox2*, *Gata3*, *Shh*, and *Lmx1a/b* null mutants for the spiral ganglion. It remains to be documented in the early spiral ganglion neurons for all tetrapods and sarcopterygians that share the *Neurog1* gene expression, also in teleosts.

In the central nervous system, distinctions between homology and analogy are not yet established with the same precision as in peripheral organs of the ear. It is necessary to show numerous analogies, necessitated by the functional constraints imposed by the auditory signal and its processing. To illustrate that point, genetic analysis has revealed in terms of previously unlikely homologies in the ear. *Lmx1a/b* is an early gene that regulates the formation of the cochlear nuclei and choroid plexus that interacts with *Gdf7* to regulate the roof plate. Without *Lmx1a/b* and *Gdf7* we see the roof plate is fused and blocks all expression of *Atoh1*, a gene needed for the entire dorsal expression of roof plates instead of a choroid plexus. In addition, we see a second expression more ventral (dA4, dB1) that expresses *Ptf1a*. There is a gradient of *Ptf1a* that is more expressed in r5 and to a certain extent r4 and is down to a sliver for r2/3. A major issue to be raised is that the data in mice needs to be expanded to other non-mammalian genes that demonstrate how similar or distinct the formation of other auditory systems are, in particular in salamanders and caecilians who have a distinct formation of central projections that retain electroreceptors and lateral line fibers (Figs. 9, 16).

In addition to these cell specific features, the evolution of molecular networks apparently generates maps of transcription factors that specify areas irrespective of their morphological details. For example, the interaction of the transcription factor orthodenticle (*Otx*) and the gastrulation homeobox gene (*Gbx*) specifies a specific area apparently in all deuterostome animals (Fig. 10). However, only chordates have developed a complex midbrain/hindbrain boundary that apparently uses this conserved transcription factor expression domain common to all triploblastic metazoans to specify a rather different morphology. *Otx* expression, together with other genes, apparently specifies the first gill slit of the spiracle, thus indicating that the area that will eventually become the auditory tympanum membrane and is one of the oldest conserved parts of the deuterostome body (Figs. 10, 27).

Based on our current insights, the vertebrate ear can be viewed as the product of continuous alteration of an existing program to govern sensory neurons and hair cell development by importing novel or existing genes to generate a novel context that alters function of those genes. Such a context dependent action has recently been reported for neuronal fate specifying genes and likely will play an even greater role in the evolution of the brain. Comparative analysis has shown a rich diversity in the position of these cells allowing only one criterion to suggest homology across taxa, the common hair cell target that will evolve into the mammalian cochlear hair cells. A strong homology is provided by the miR-genes (miR-96, 182, 183) that define all sensory hair cells, including mechanosensory hair cells of the lateral line, the vestibular hair cells, the electroreception hair cells, and the auditory hair cells.

Vertebrate ears, including both vestibular and auditory endorgans, are sensitive to particle motion. These projected sensory neurons reach to a ventral octaval column. This system was transformed by the evolution of sound pressure-sensitive receivers, and by the evolution of a dorsal acoustic column that receives central auditory targets. The task of relating the evolution of sound pressure receivers to the evolution of the central auditory system requires resolving the evolution of three interlinked developmental steps:

1. Generate sensory neurons that interconnect this new auditory endorgan to an area of the brainstem dedicated to sound pressure reception that must have evolved.
2. Evolving auditory nuclei *de novo*, or from some undifferentiated precursors, or respecify existing other targets throughout the CNS of vertebrates (lateral line, electroreceptors) to form targets (auditory nuclei) specifically dedicated to processing sound pressure signals.
3. Generate a redundant endorgan not required for vestibular responses and so can be modified to extract sound pressure from various stimuli impinging on the ear.

It comes as no surprise that ideas of transformational identities, like the evolutionary transformation of the hyomandibular bone into the stapes of the terrestrial ear (Figs. 23-28) and proposed that auditory nuclei derive from the mechanosensory lateral line and electroreceptor areas (Fig. 9), a suggestion rooted in the limited resolution of early neuroanatomical tracing studies. It was further suggested that auditory input from the ear may be replaced, evolutionarily as well as during the metamorphosis of frogs, the degenerated lateral line input. Although this idea is interesting, the main problem is that not all frogs lose the lateral line system during metamorphosis, and even those that retain the lateral line system have auditory nuclei themselves: *Xenopus* is a frog that retains the lateral line system and has an auditory system. Specialized auditory nuclei are found even in some bony fish, which fully retain the mechanosensory lateral line system. Thus, neither the apparent coincidence of loss and appearance of another nucleus nor the detailed connections support this idea of transformational identity. The only factors supporting it are more generalized connections to the midbrain, the inferior colliculus (IC) in mammals, called torus semicircularis in frogs and other non-mammals.

More recent data suggest yet another transformation, related to the discovery that an electroreceptive sense is primitive for all jawed vertebrates, including lampreys and basic gnathostomes, including certain amphibians. Loss of this primitive sense of electroreception fits the formation of specific auditory nuclei correlate among frogs and amniotes. In contrast to the mechanosensory lateral line nuclei, electroreceptive and auditory nuclei show topological similarity in their rostrocaudal extent in mammals (from trigeminal to glossopharyngeal roots or from rhombomere 2 to 6), whereas lateral line and vestibular nuclei extend from rhombomere 1 to rhombomere 7/8, including the chicken auditory that extends past r6. Moreover, among salamanders and caecilians, some have electroreception associated with the specialized CNS nuclei, but no specialized auditory nuclei are known (salamanders, caecilians). Thus, at least a spatial correlation exists between the loss of the ancient sense of passive electroreception and the gain of auditory nuclei in the brainstem with the overall topology of the tetrapod auditory nuclei, including frogs and amniotes. How the molecular expression of the most dorsal auditory nuclei must have transformed into the auditory system in frogs and tetrapods is unclear.

In summary, we need to understand the auditory system in all sarcopterygians that receive a unique auditory neuron to reach out to the auditory nuclei and connect the neurons with the auditory hair cells.

4.1. Auditory spiral ganglion neurons are unique among mammals

Sensory neurons connect spiral ganglion neurons with the peripheral hair cells and central cochlear nuclei. A unique development starts in *Latimeria*,

Defining the Auditory System of Tetrapods 117

a sarcopterygian, which has a separate branch to innervate next to the lagena, the basilar papilla. A set of unique genes characterized in detail that loses the auditory ganglion neurons entirely or in parts. Unfortunately, data are mostly restricted to mice mutations for which we have the spiral ganglion neurons.

Vestibular ganglion neurons (VGN) and spiral ganglions neurons (SGN) are independently derived from a common origin, *Eya1*, *Sox2*, and *Neurog1* (Fig. 29). *Eya1* and *Sox2* define SGN precursor populations whereas *Neurog1* is needed to initiate the proliferation and differentiation of SGNs. Without *Eya1*, *Sox2*, or *Neurog1* no sensory neurons will develop. Downstream there is a segregation of genes that are positive for *Tlx3* for vestibular neurons but are negative for SGN. Furthermore, *Sall3* is in part positive for certain VGNs that are negative for other VGNs. The diversity of two types of SGNs can be defined: *Neurod1, bHLHe22, Pou4f1, Runx1, Prox1*, and *Gata3* determines the type 1c (Fig. 29). The expressions of *Tshz2, Grhl1, Rxrg, Tie4*, and *Id1* are needed for type Ia (*Runx1*), Ib (*Gata3*) and type II (*Etv4, Prrx2*) SGNs. In the end, we have the upregulation of *Pou4f1, Lypd1* and *Mgat4c* (Type Ic), *Lypd1, Pou4f1* and *Calb1/2* (Type Ib), *Calb1*,

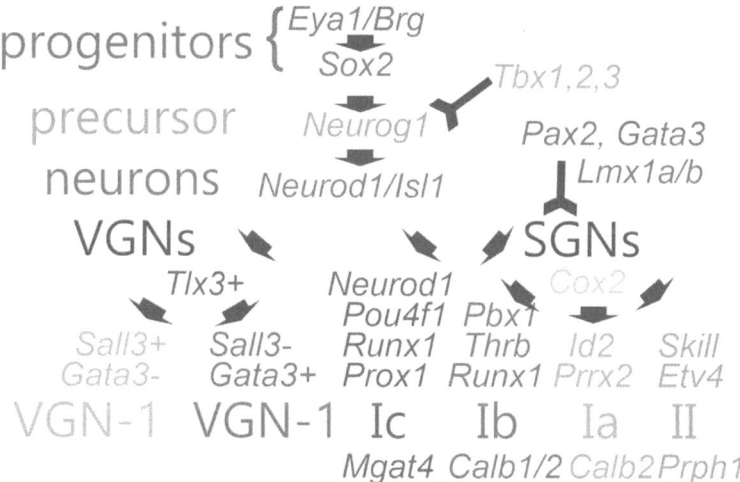

Figure 29: Progenitors of mammals are dependent by initial *Eya1*, *Sox2* and *Neurog1* that are needed for several genes to differentiate neurons. Downstream are a large set of genes, among them are *Neurod1*, *Isl1* and *Pou4f1*. A split is dependent on *Tlx3* that form VGNs whereas SGNs are not dependent on *Tlx3*. After that the upregulation of *Sall3* is needed to make two different kinds of VGNs. SGNs split into four populations, Ia, Ib, Ic, and II. A unique set of genes and their expression needs to be verified to develop each type of SGNs, notably *Pax2*, *Gata3* and *Lmx1a/b*. Image is compiled from (Dvorakova et al., 2020; Elliott et al., 2021a; Filova et al., 2022b; Kaiser et al., 2021; Petitpré et al., 2022; Sanders and Kelley, 2022; Sun et al., 2022; Xu et al., 2021).

Pcdg20 and *Prph1* (Type Ia) and *Plk5*, *Prph1* and *Th* [Type II (Elliott et al., 2021a; Petitpré et al., 2022)]. In addition, we are to deal with the *Nhlh1/2* and *Isl1* that are interacting with *Neurod1* to regulate SGNs that develop VG neuron formation.

Several selective genes can result with losses of SGNs. Conditional deletion of *Gata3*, *Lmx1a/b*, and *Dicer* as well as deletion of *Pax2* results in complete loss of SGNs (Fig. 30) while many vestibular neurons develop despite the loss of SGNs. Deletion of the gene *Dicer* (Fig. 30A), which is needed to make microRNA (miR), is blocking all cochlear hair cells and spiral ganglion neurons that receive very few vestibular utricular hair cells that receive a few vestibular neurons. The cochlea is reduced into a sac consisting of the reduction of the lagena in specific amphibians.

Gata3 provides a unique molecular identifier of auditory sensory neurons (Fig. 29). In the absence of a cochlear formation, we find a sac that is devoid of cochlear hair cells and cochlear neurons (Fig. 30B). Even if part of *Gata3* vestibular neurons is forming, such as *Pax2-cre; Gata3-/-*, a few patches of SGNs and hair cells that receive a pattern of innervation that reaches the fibers to the cochlea.

Pax2 is among the earliest gene investigations that lack the formation of hair cells that form a sac (Fig. 30C). The position of the sac is inside the brain below the brainstem where they attach. Neurons form initially but undergo a rapid loss. Obviously, central projections from the cochlea are absent. Vestibular neurons and vestibular hair cells form. However, without *Pax2* and *Pax8*, the ear does not develop beyond an otocyst comparable to *Eya1* and *Sox2* deletions.

Lmx1a/b null mice show a loss of cochlear hair cells, cochlear neurons, and cochlear nuclei. Without *Lmx1a/b* double null mutants we have no formation of cochlear hair cells that form vestibular hair cells with some changes. Likewise, the cochlear neurons are missing that show a

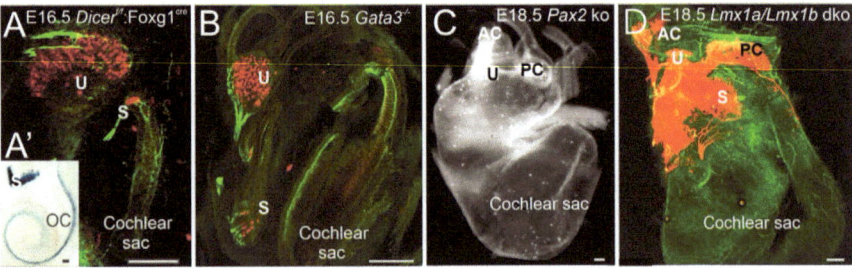

Figure 30: This image shows the loss of cochlear sensory cells in all four mutant mice that have a variable expansion referred to as a sac (A-D). All four cases also lack spiral ganglia neurons. Insert (A') shows the hair cells of the cochlea with *Atoh1-LacZ*. Note the variable development of vestibular hair cells (A, B) and vestibular afferents (D) are shown. Image is provided by (Fritzsch, 2021; Fritzsch et al., 2013).

prominent expansion of the cochlear sac (Fig. 30D). In addition, we know that cochlear nuclei are not forming in *Lmx1a/b* double null mice that show a loss of the choroid plexus and *Atoh1* that defines the dorsal most expression in the brainstem.

Unfortunately, the expression of *Lmx1a/b* is required for the dorsal part of the hindbrain, which has not been analyzed in the lateral line and electroreception in gnathostomes. It is possible that the lateral line and electroreception may play a role in *Lmx1a/b* expression to help the transformation of amniotes after the loss of peripheral hair cells and associated nuclei and central projections. Recent evidence has shown that cyclostomes have a different organization of *Lmx*, but their expression of *Lmx1a/b* is unclear. Moreover, the two groups of teleosts that have evolved electroreception have a unique expansion among all gnathostomes. This expansion mimics the auditory system of amniotes, for which information on *Lmx1a/b* expression is lacking.

A unique set of *Tbx1,2* and *3* deletions interact with *Neurog1* to downturn VGs and SGNS (Kaiser et al., 2021). *Tbx1* acts as a selector gene that controls neuronal fate in the otocyst. Ablation of *Tbx2* led to cochlear reduction but it is unclear how many spiral ganglia neurons form. Likewise, the absence of *Tbx3* results in vestibular malformations that require the distribution of vestibular ganglion neurons. Combined deletion of both *Tbx2/3DKO* results in incomplete and largely absence of cochlear and vestibular structures in newborn mice (Fig. 29) for which we have no presence of vestibular ganglion and spiral ganglion neurons. A triple deletion of all three *Tbx* genes will likely result in *Eya1, Sox2, Neurog1,* or *Pax2/8* deletions that lack all VGNs and SGNs.

Another gene that is needed for normal cochlear neurons is the sonic hedgehog (*Shh*). In the absence of *Shh,* it results in the absence of a cochlea and the loss of SGNs. An interesting expression of SGNs in the earliest *Shh* dependency but shows limited changes of cochlear hair cell interactions (Elliott et al., 2021c).

Downstream are neurotrophins that are needed for their sensory neuron development and maturation. The loss of all SGNs is documented for *TrkB* and *TrkC* that signal for *Bdnf* (*TrkB*) and *Ntf3* (*TrkC*). There is an incomplete deletion after the basal turn that is dependent on *Ntf3* and *TrkC* while the loss of *Bdnf* and *TrkB* causes the reduction and loss of the apex SGNs: *Bdnf* depends on 95% of vestibular ganglion neurons while only about 5% are lost in the SGN, mainly in the apex. In contrast, *Ntf3* (*NT3*) nearly lost 95% of SGNs, and has lost all basal turn into SGNs, but has only an additional loss of about 5% of VGNs (Elliott et al., 2021b).

Proliferation forms from E9-14 for VGs whereas SGNS are proliferation from E10.5-12.5. VG neurons delaminate and migrate to form a ganglion outside of the ear while the SGNs stay inside the cochlea. Developing neurons reach out first from VGNs at E10 whereas the first

SGNs fibers reach the auditory nuclei at E12.5. Certain deletions of genes (*Neurod1, Isl1*) as well as loss of Schwann cells after *Sox10* or *ErbB2* show a migration of SGNs to mix with the VGNs. There is a delay between SGNs and cochlear hair cells: SGNs form first in the base (E10.5) to the apex (E12.5). In contrast, cochlear HCs start from the apex around E12.5 and progress to the base about E14.5. *Atoh1* is needed for all HCs but has a different progression starting in the base at E14.5 and progressing to the apex around E18.5. Innervation is reaching IHCs starting at E15 that expand to reach OHC innervation by Type II fibers at E18 (Fig. 31).

Innervation forms from the brainstem to reach out selectively to inner ear efferents (IEE) to innervate the cochlea hair cells. The lateral olivocochlear bundle (LOC) comes from mostly ipsilateral innervation. The innervation originates from mostly the contralateral origin that reaches out as the medial olivocochlear bundle (MOC) to innervate the OHCs (Fig. 31).

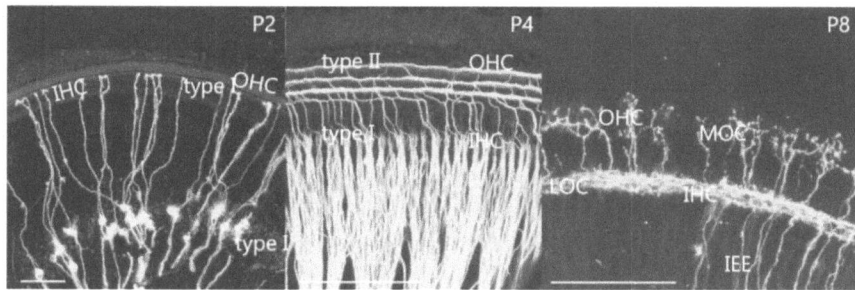

Figure 31: Labeling from r7 injections of dye from the brainstem can trace the type I neurons that innervate selectively a few SGNs that innervate the IHC. HC are autofluorescent (lilac) (light gray). *Prph*-eGFP are positive for all type I neurons to the IHC and have type II projections to OHCs. Inner ear efferents (IEE) are labeled with dye injections that form two bundles: the lateral olivocochlear (LOC) ramify among IHC innervation whereas the medial olivocochlear (MOC) innervation reaches out the OHC. Images compiled from (Elliott et al., 2021a; Fritzsch et al., 2010; Zuo et al., 1999).

Older mice and humans have lost progressively the SGNs, starting near the base (Elliott et al., 2022b; Fritzsch et al., 2022). Fortunately, many neurons survive in the absence of hair cells that can be used to drive as a cochlear implant to excite cochlear nuclei. The absence of hair cells result in a flat epithelium for which the hope is to eventually restore lost hair cells to connect with the surviving SGNs.

In summary, a set of genes is needed to govern the formation of the SGNs and depends on their development that diversifies into four SGN types. Future work is needed to define the role of the distinct formation of SGNs that innervate the cochlear nuclei and the HCs. We know that

Defining the Auditory System of Tetrapods 121

distinct efferent fibers develop into IHCs and OHCs that provide a functional connection directly to OHCs and indirectly to SGN fibers near IHCs. An overview of connections will be provided below in 4.2 and 4.3.

4.2. Auditory nuclei are expanded to add additional connections

Connections have an auditory system that is added to the brainstem nuclei (auditory nuclei, superior olivary complex, lemniscal nuclei) that are not recognized in the electroreceptive or mechanosensory lateral line systems (Fig. 32). Instead, a direct connection seems to be a major projection to the torus semicircularis in either lateral line or electroreception. It needs to be

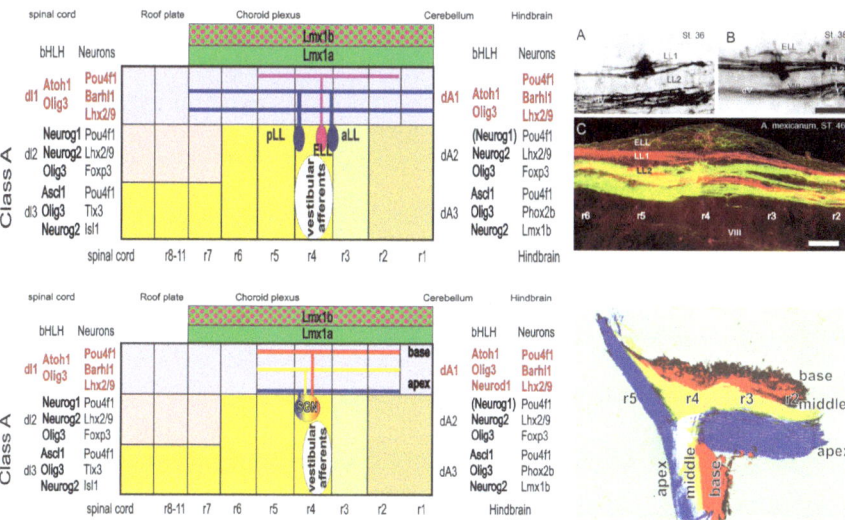

Figure 32: Comparing the organization of electroreception and lateral line fibers (top) with the mammalian auditory nuclei (bottom). *Atoh1* interacts with *Olig3* and *Neurod1* to drive the spinal cord, brainstem, and cerebellum. Within that common expression, certain areas are forming. Lateral line develops first in salamanders that later receives electroreception fibers. A few fibers from the auditory nerves are overlapping with salamanders with the vestibular nuclei which contributes a large area from several future nuclei. Mammals develop in a base-to-apex progression, inverted to the progression of lateral line-to-electroreceptors (top and bottom). Additional contribution to *Atoh1* provided by *Ptf1a* is from dA4/dB1 and dA1 (*Atoh1*). How much contribute for transforming the electroreception/lateral line of salamanders and caecilians into genes to contribute the auditory nuclei is unclear for mammals nor is the contribution from *Ptf1a* in amphibians. What is obvious is the similar rostro-caudal extend from r2-5 in salamanders and mammals. Modified after (Elliott et al., 2021c; Fritzsch, 2021; Macova et al., 2019).

pointed out that the novel electroreceptive sense of derived teleost fish seems to have evolved anew in a pattern grossly similar but not identical to the ancestral passive electroreceptive sense, after this sense was apparently lost in ancestral teleost fishes. In fact, among teleosts, some special auditory systems are well known in mormyrids and silurids, among others. If the auditory nuclei of tetrapods represent a comparable transformation of a dormant developmental program, will be called novel connections. Such functional reconnections have been experimentally shown for forebrain connections and certainly is a possibility worth exploring. Unfortunately, no recent molecular evidence is provided in other teleosts that focus on fish that do not develop electroreception. However, the zebrafish is well known that they have developed an auditory system in addition to the lateral line.

Attempts to experimentally reorganize the lateral line or auditory input by transplanting the ear or extirpating the ear in frogs showed little alteration in the remaining projections. It also needs to be stressed that this entire idea is not yet tested on the molecular level: no specific lateral line, electroreceptors, or auditory nuclei markers are known that highlight all sensory nuclei of each of these systems across taxa independently. More recent molecular data, proposes that functional systems may be developmentally connected by shared activation of transcription factors, suggesting that such molecules may exist. Moreover, auditory nuclei appear to express unique markers in some songbirds. It is therefore possible that identical transcription factors govern the development of both the peripheral receptors (i.e., sensory neurons, sensory cells) and central nuclei (i.e., lateral line, electroreceptive, and auditory nuclei in the brainstem). If this can be demonstrated, it would appear more plausible that all parts of these systems are lost concomitantly during evolution, presumably owing to a single or few mutations.

If proven, this would eliminate the core of the replacement hypothesis and both the auditory periphery, and the functionally related auditory nuclei would then need to be considered as evolutionary novelties that, nevertheless, might represent transformations of a general column of cells extending throughout the hindbrain and spinal cord. In the following discussion, the formation of pressure-sensitive organs in the auditory periphery and central nuclei is an evolutionary novelty, starting with the evolution of lungs and tympanic membrane. Logic requires that the formation of specific endorgans predates that of specific auditory nuclei (one needs to extract auditory information before it can be processed), that is, the evolution of the auditory system needs to be resolved in an ear-to-brain progression. The evolution of the inner ear will develop first, followed by the limited insights into the formation of endorgans specific connections and the development of auditory nuclei. Auditory sound pressure receptors of the ear are uniformly characterized by otoconia-free tectorial membranes and do not respond to gravistatic or angular acceleration stimuli.

A second, unifying feature of sound pressure receivers is the association of the sensory epithelium with a system that allows sound pressure to induce fluid motion only in the vicinity of this sound pressure-receiving organ. In contrast to the multitude of apparently parallel transformations of gravistatic receptors into sound pressure-receiving auditory endorgans among teleost fish and the unique amphibian papilla of amphibians, there is apparent uniformity in topology and innervation of the main sound pressure-receiving endorgans of tetrapods, the basilar papilla. Development in amphibians and mammals suggests that basilar papilla/cochlea become progressively segregated from a common precursor, suggesting that the basilar papilla evolved only once among the aquatic ancestors of tetrapods, the sarcopterygians shown in *Latimeria*.

Irrespective of their peripheral distribution or their embryonic origin, all sound pressure receivers appear to project discrete nuclei of the brainstem that receive only limited input from other inner ear endorgans. This requires that sound pressure receiver afferents carry specific markers that allow them to project differently from the nearby vestibular afferents. In mammals, such a differential projection develops prior to the formation of sensory hair cells and thus cannot be mediated by hair cell activity. Indeed, topologically correct projections of the auditory system, including the tonotopic organization of afferents, develop before birth and even in the absence of hair cells. Most recent data have identified a set of genes (*Atoh1, Ptf1a*) that is uniquely expressed in cochlear nuclei and is, in other rostro-caudal expression, involved in pathway selection.

In summary, recent ideas about the transformation of lateral line nuclei into auditory nuclei are doubtful given the co-existence of lateral line and auditory nuclei in *Xenopus*. One possibility is that the entire anlage of lateral line and electroreceptive nuclei (or just the electroreceptor nuclei) has been transformed by the expression of another set of genes of *Atoh1* into auditory nuclei in amniotic vertebrates. A set of genes is required to be investigated such as *Lmx1a/b, Gdf7, Atoh1,* and *Ptf1a*, among others, may eventually be investigated for lateral line and electroreceptors to generate auditory nuclei instead. Understanding how novel auditory nuclei arise in a network of highly conserved transcription factors that predates the evolution of the vertebrate brain and that provides a blueprint for topographically specific neuronal development is the most challenging and yet least understood aspect of brain evolution. I will deal with this aspect in Chapter 5.

4.3. A unique set of genes are needed for hair cells in mammals

Cochlear hair cells may be traced to the origin of ancestral *Latimeria* in which we have a basilar papilla in most sarcopterygians, except for lungfish and derived amphibians that have lost the basilar papilla and hair cells. All amniotes have a basilar papilla that transforms into mammals and becomes the organ of Corti. A notable transition is found in monotremes that have a basilar papilla next to the lagena that shows already the two types of hair cells, 2-4 inner hair cells and up to 12 or more outer hair cells. In contrast, mammals have lost a lagena and have a single row of inner hair cells and three rows of outer hair cells.

With respect to SGNs for the mammalian cochlea, a unique and common gene expression is needed for developing the cochlear hair cells, the inner hair cells (IHCs), and the outer hair cells (OHCs). Upstream is the expression of *Eya1* and *Brg1* that are needed for the initiation of hair cell progenitors. Downstream is the expression of *Sox2* to interact with bHLH genes to drive the cochlear hair cells to initiate hair cells. The bHLH gene *Atoh1* (Fig. 32) is needed for vestibular and cochlear hair cell formation beyond undifferentiated hair cells. Downstream are *Pou4f3*, *Gfi1* and *Barhl1*. Differential effects between OHC and IHC are partially characterized.

An interaction is existing between the bHLH gene *Neurog1* and *Foxg1* that causes a reduced expression of *Atoh1*: The absence of *Neurog1* or *Foxg1* result in a short and wider cochlear set of hair cells that results in more rows of outer hair cells (instead of three rows of outer hair cells). Misexpression of *Neurog1* instead of *Atoh1* develop IHCs and OHCs that are not functional and are converted by inner pillar cells into gaps between near normal IHCs. An interaction also shows a shorter cochlea and converts OHCs into IHCs after the deletion of *Neurod1*, a bHLH gene. The proliferation of more hair cells depends on n-*Myc* and will transform into the apex akin to a lagena-like organization.

For example, *Emx2* reduces all OHCs and has two or more rows of IHCs while the loss of *Fgf20* results in two rows of OHCs. Downstream are *Insm1* and *Ikzf2* which are needed for OHC development; OHC development interacts with *Neurod1* which also regulates OHCs instead converted to IHC. In contrast, loss of *Tbx2* converts IHC into OHC-like hair cells (García-Añoveros et al., 2022). Downstream of IHC is *Fgf8* and *Srrm3/4* (Nakano et al., 2020) which will lose nearly all IHCs in the absence of two selectively expressed genes (Fig. 33).

Several genes are required for cochlear development which results in the absence of all cochlear hair cells. *Gata3*, *Lmx1a/b*, *Pax2*, and *Dicer* (Fig. 33) do not develop any hair cell that likely interacts with *Atoh1* for hair cell

Defining the Auditory System of Tetrapods 125

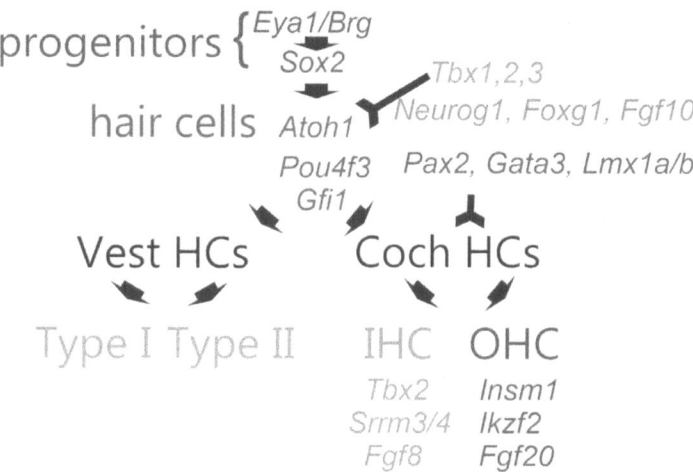

Figure 33: Progenitors are followed by *Eya1*, *Sox2* and *Atoh1* that are needed for hair cell developing. Downstream are *Pou4f3* and *Gfi1* that are needed to maintain hair cells. *Tbx1,2,3* interacts with *Neurog1* and *Foxg1* regulates the number and distribution of cochlear to make it shorter and has increased the number of hair cells. A loss of all cochlear hair cells is revealed in *Pax2*, *Gata3* and *Lmx1a/b* mice. Two types of vestibular hair cells exist that have a mixed distribution of type I and type II HCs. In contrast in the cochlea, we have a single row of IHCs and three rows of OHCs. *Tbx2*, *Srrm3/4* and *Fgf* are needed for differentiation and viability of IHC. *Insm1*, *Ikzf2* and *Fgf20* are needed to differentiate OHCs or requires for forming three rows of OHCs. Image is compiled from (Chizhikov et al., 2021; Filova et al., 2022a; García-Añoveros et al., 2022; Kaiser et al., 2021; Matei et al., 2005; Nakano et al., 2020; Pauley et al., 2006; Xu et al., 2021).

formation (Chizhikov et al., 2021). In addition, we have a delayed loss of all hair cells that initially develop in a loss of *Bdnf/Ntf3* double deletions (Kersigo and Fritzsch, 2015). Moreover, the loss of *Bdnf* deletion will result in the apical OHCs (Fig. 34) while the loss of *Ntf3* cause the loss of OHC in the base. Likewise, it is dependent and will degenerate in *Cdc42* for IHC first, followed by OHCs in a base to apex progression. In addition, *MANF* deletion results in the OHCs in the base. In contrast to this complete loss of cochlear hair cells, many vestibular hair cells develop in the absence of these genes.

Hair cells started with a single kinocilium which is surrounded by microvilli in choanoflagellates that is an initial developmental step in mammals (Fig. 6). We know a large set of genes are needed to develop the hair cell tip from a symmetric to an asymmetric distribution. In addition, stereocilia elongate closer to the kinocilium that can form two hair cell types with respect to stereocilia: thinner stereocilia develop into type II vestibular hair cells and the OHC whereas the thicker stereocilia become type I vestibular hair cells and the IHCs. The transient kinocilia are lost

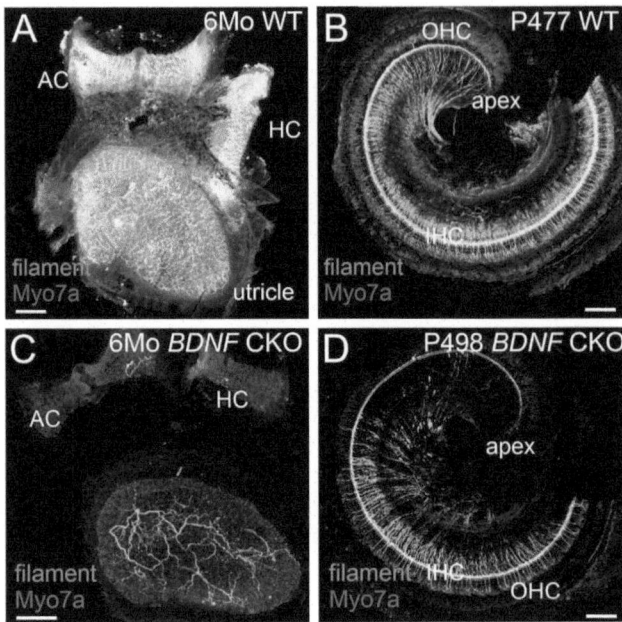

Figure 34: The effect of loss of *Bdnf* resulting in the loss of SGNS and OHCs in older mice (D) compared to littermates (B). In contrast, *Bdnf* shows the near complete loss of vestibular neurons that show a minor reduction of HCs (C) compared to control littermates that show a profound innervation using filament immunocytochemistry. Modified after (Elliott et al., 2021b) and unpublished.

before the function of the auditory starts, whereas the vestibular hair cells maintain a kinocilium.

Inner and outer hair cells differ in the organization of stereocilia: both types develop after the kinocilia is retracted but have two rows of thick IHC and two-three rows of thinner OHCs, both polarized toward the stria vascularis. The tip links depend on *Cdh23* at the taller stereocilia and *Pcdh15* is with the shorter stereocilia that connects with *Tmc1*, *TMIE*, and *Cib2* and others (TMEM16, and possibly *Piezo*) for opening the potassium channels if the stereocilia are deflected toward the stria. The detailed multimers of proteins are incompletely understood that and each deletion results in the inability to open the channel (*Cdh23*, *Pcdh15*, *Tmc1*, *TMIE*, *Cib2*). As stated before (Fig. 2), the detailed analysis is open to describe the connections for that allow the MET to open it up. It relates to the tectorial membrane that has a direct connection with the OHCs using Prestin to contract the OHCs but is freely moving within the fluid with the IHCs.

The organ of Corti has unique features for inner and outer hair cells that differ in their function (Fig. 35). The flow between the inner spiral sulcus and the tectorial membrane allows the stimulation from stereocilia. In addition, a contractile system is affecting the OHCs. An interesting ratio is depicted by IHC/OHC that is nearly 4:5, making many more of each three rows of OHC compared to the single row of IHCs (Fig. 35). The distribution of many more IPCs is forming next to OPCs (a ratio of 12:9).

Age related hearing loss is a significant effect that will affect close to 1.5 billion people over 65 years of age (Haile et al., 2021): hearing loss starts with the high frequency loss that loses nearly all basal turn of OHCs followed by IHC reduction and loss. Over 300 genes are associated with hearing loss (Elliott et al., 2022b). Attempting to replace the lost hair cells in older people and mice would be helpful, but we have no idea how to start novel hair cells. Obviously, an interaction with *Sox2, Atoh1, Pou4f3,* and other genes are needed to initiate and regulate their expression to restore hair cells.

A continuation has been described in detail (Michalski and Petit, 2019; Moser, 2020). Inner hair cells (IHCs) using Otoferlin (*Otof*) function for temporally sound that encodes their afferent synapses of the SGNs, the three types of SGNs. Ribbons provide the presynaptic active zone that interacts with the voltage-gated channels. Synaptic vesicles are

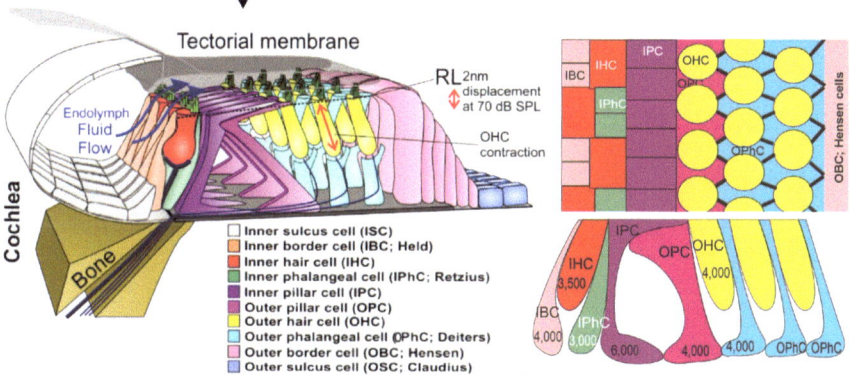

Figure 35: (Left) The organ of Corti has a unique organization of hair cells and supporting cells to form one row of IHCs and three rows of OHCs. Three types of supporting cells form: the inner border cell (IBC) and inner phalangeal cells (IPhC); the inner and outer pillar cells (IPC, OPC) and the outer phalangeal cells (OPhC; Deiters) and the outer border cells (OBC; Hensen). Following a downward deflection of the reticular lamina (RL) by 2 nm at 70 dB sound pressure level (SPL). Endolymph flows between the inner spiral sulcus and excites the stereocilia as they pass between the tectorial membrane and the IHC/OHC stereocilia. OHC has a unique contraction upon proper stimuli. (Right) The numbers of cell types are different, for example, a ratio of nearly 1:2 between IHC and IPC whereas a ratio of 1:1 exists between OHC and OPC. An interesting difference is shown between IPC and OPC (12:9). Modified after (Elliott et al., 2018; Jahan et al., 2015a).

cooperating with the ribbon to reach the active zone to allow Ca^{2+} to release the membrane. Ribbons that dock the synaptic vesicles allow them to reach the active zone to release the content (Ca^{2+}, neurotrophins) into the cleft to be diffuse to interact with afferent SGNs fibers. The extraordinary function of sound encoding has a very high rate of transmitter release per active zone. A somewhat similar vesicular dependency is less well understood for the OHCs that seem not to contribute to sound encoding. Instead, OHC has a direct input of efferents, whereas IHC interacts with the SGN afferents that require detailed analysis for their function.

In summary, cochlear hair cells are distinct in mammals that evolved from the monotreme organization. How the evolution of the basilar papilla, that Corti, eventually becomes the organ of Corti requires further investigations of the hair cells in other sarcopterygians (*Latimeria*, amphibians, and amniotes), including the monotremes, that will define molecularly the origin of the cochlear hair cells of mammals.

CHAPTER 5

Sound Processing: Boundaries for Auditory Signal Processing Revealed

The auditory system is known among bony fish, amphibians, sauropsids, and mammals (Carr, 2020; Grothe et al., 2004). Bony fishes develop a separate way to generate an auditory system that does not require the tympanic membrane but reaches the auditory input through various means to direct or indirectly drive the lungs to reach various sounds, mostly through the saccule but have connections from the utricle (Schulz-Mirbach et al., 2019). Sound perception is driven by an extension of the lungs to the ear in otophysans, clupeids, and gnathostomes. Other bony fish have a Weberian system to drive the lungs indirectly. Unfortunately, limited information about sound perception is unclear in the polypterus which has a lung without any known input to reach the ear. Evidence shows that most bony fish can respond to sound for mostly low frequencies that show increasing sensitivity to reach below 1 kHz with special connections to up to 4 kHz or more (clupeids, silurids). Compared to other teleosts, lampreys, and lungfish have a lower sensitivity of around 200-300 Hz. A very large specific input of the elasmobranchs is described by a papilla neglecta that has some 200,000 hair cells. Sound responses are usually around 300 Hz and can reach up to 1 kHz or slightly more. How the sound is perceived without a lung is unclear in lampreys, elasmobranchs, and lungfish, the latter has no connection between the ear and the lungs. In contrast, sound depends on the lungs, and the tympanic membrane and depends on the basilar papilla, tectorial membrane, and round window that evolves into the cochlea (Chapter 3). The next step is describing the central auditory nuclei that depend on all tetrapods for which we have a limited central projection in *Latimeria*, most caecilians and salamanders, and several central projections in most sauropsids and monotremes.

5.1. Auditory neurons innervate selectively the topologically auditory nuclei

Spiral ganglion neurons develop between E12-14 in a base to apex progression in mice. Some misunderstanding occurs about the earliest incomplete overlap from basal SGNs with delaminating saccular neurons that migrate to join with other vestibular neurons. Unfortunately, the use of cochleovestibular ganglion (CVG) is a misnomer that does not fit the cochlear projections which are segregated for the central projection. An overlap between the SGNs and VGN can happen as additional migrations occur to generate a CG in mutants such as *ErbB2, Sox10, Neurod1,* and *Isl1* (Filova et al., 2022b). It is noticeable that the first central projection of cochlear neurons develops to reach the cochlear nuclei at E12 and is exclusively derived from basal turn afferents (Schmidt and Fritzsch, 2019). As we progress the SGN central projection, we find that a topological innervation develops between the basal turn (dorsal) and the apical turn (ventral) to have the topology of afferents established at E14. A few fibers from the saccule can reach out in the most ventral projection (Fig. 36).

SGN afferents are topologically organized to the AVCN but deviate from the DCN (Fig. 37) where it reaches out the organization in the adult mice (Muniak et al., 2016). Analyzing the central projection of SGNs to reach out to the cochlear nuclei shows distinct losses and reorganization of the tonotopic organization. The incomplete loss of central projections can cause the reorganization and end up with a mixed innervation from the SGNs either *Neurod1* or *Isl1* that results in fewer SGN neurons that migrate to mix with VGs (Filova et al., 2022a; Filova et al., 2020; Filova et al., 2022b; Macova et al., 2019): Many deletion of either *Neurod1* or *Isl1* results in a central overlap of basal and apical projection that shows an incomplete and mostly overlapping of much fewer SGN fibers that show the massive loss and reorganization of cochlear fibers (Fig. 37). Ephrins, particularly *Ephrin-A3*, are playing a role that causes an expansion of cochlea nuclei projections (Hoshino et al., 2021; Krasewicz and Yu, 2022) consistent with the loss of *Neurod1* and *Isl1* deletions. In addition, tonotopic gradients of VGLUT1, VGAT, Syt1, and Syt2 play a role with higher protein densities in the more medial (high-frequency) area (Yu and Wang, 2022). Since *Atoh1* is needed for cochlear nuclei development, one can investigate the deletion of all *Atoh1*, selectively deletion of only the cochlear hair cells, or incomplete deletion of either *Hoxb1* or *Krox20* (*Egr2*). Since SGNs project near normal using *Isl1-cre; Atoh1* conditional deletion in cochlear hair cells develop near normal SGNs at E14.5 that degenerate rapidly in the absence of cochlear hair cells. The tonotopic organization (Fig. 37) is near normal after either *Atoh1* cochlear hair cells after conditional deletion of *Isl1* (Filova et al., 2020) or even after the loss of *Atoh1* in cochlear hair cells

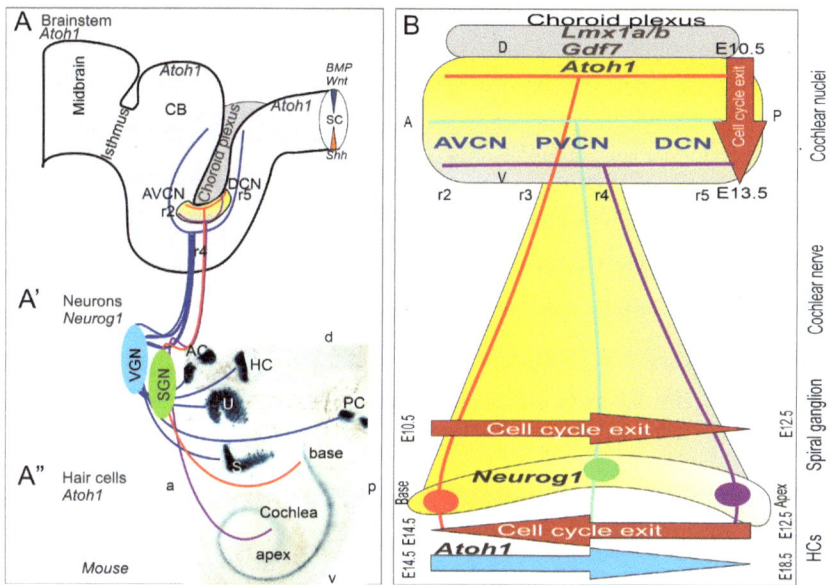

Figure 36: Cochlear and auditory projection development depend on the expression of specific genes. The development of spiral ganglion neurons (SGN; A') depends on the expression of *Neurog1*. Without *Neurog1* no vestibular or cochlear neurons develop, their projection extends from the cochlea first (A", B) and ends in a topologically organized projection in the auditory nuclei (AVCN, PVCN, DCN). The development of auditory nuclei depends on the expression of *Atoh1* (A, B). Also shown are the vestibular ganglion neurons (VGN, A'), which project from the 5 vestibular sensory endorgans (anterior, posterior and horizontal semicircular canals, AC, PC, HC, and utricle and saccule, U, S; A") to the central vestibular nuclei and cerebellum (CB; A). SGNs proliferate in a spatiotemporal gradient from the base to the apex of the cochlea during the embryonic period E10.5-E12.5 (B). Their central projections develop topologically from dorsal to ventral positions within the central auditory nuclei over approximately the same period (E10.5-E13.5; B). Later, hair cells proliferate in a gradient from the apex to the base of the cochlea during the period E12.5-E14.5 (B) and contact the peripheral terminals of the SGNs. *Atoh1* expresses after the proliferation is completed that it progresses from E14.5 (base) to reach the apex as late as E18.5 (apex; insert in A, *Atoh1-LacZ*). AC, anterior crista; AVCN, anteroventral cochlear neurons; CB, cerebellum; DCN, dorsal cochlear neurons; HC, horizontal crista; PC, posterior crista; r2/4/6, rhombomeres; S, saccule; SC, spinal cord; U, utricle. Modified from (Filova et al., 2022b; Fritzsch et al., 2019; Glover and Fritzsch, 2022; Macova et al., 2019; Nichols et al., 2008).

and cochlear nuclei (Elliott et al., 2017). In addition, the loss of *Nrp2* results in shorter rostral and caudal extension that is much broader compared to control mice (Schmidt and Fritzsch, 2019).

Deletion can affect to a certain extent the tonotopic organization after a conditional deletion of *Hoxb1* of r4 or *Egr2* in r3/5, in particular using *Atoh1*CKO deletion (Di Bonito and Studer, 2017; Maricich et al., 2009). Obviously, after deletion in r4 or 3/5 end up with a shorter rostro-caudal extension compared to control mice (Fig. 37): the more profound projection of r3/5 deletions of *Egr2-cre; Atoh1*DKO reaches more DCN fibers whereas the AVCN is ramifying beyond the r2 terminals. Likewise, a shorter rostral projection will be reduced in r4 *Hoxb1-cre; Atoh1*DKO for the AVCN but has a near normal DCN fiber projections (Fig. 37). A noticeable projection form on *Egr2; Atoh1*DKO while the anterior projection is shorter and disoriented. It is possible that the more profond caudal projection could correlate with the much more profound neurons that derive from *Ptf1a* that should be normal after *Atoh1* deletions (Fujiyama et al., 2009). A conversion using *Atoh1-cre* to use the expression of *Hox2* (*Hoxa2* and

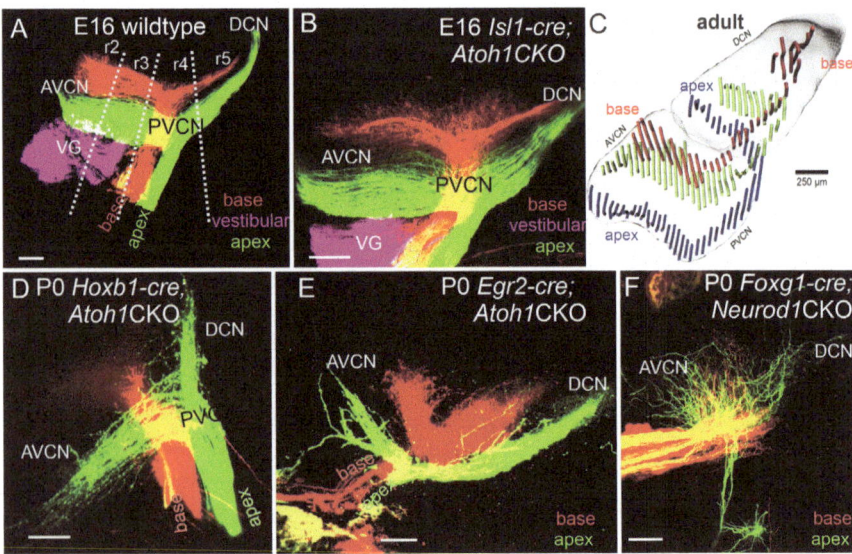

Figure 37: The central projection of SGNs reaches the cochlear nuclei by E14 and follows a tonotopic organization in control mice (A) and has a near normal tonotopic projection in the absence of cochlear hair cells (B). The tonotopic organization is presented for adult mice (C) that show a different projection from the DCN where the fibers interact between the base and the apex. Cochlear projections are shortened due to the loss after the loss of r4 in *Hoxb1* (D) which is even shorter in the r3/5 deletion of *Egr2* (E). Note that the SGNs are reduced but shows a significant loss of *Egr2* deletions that has a more profound DCN compared to ramifying AVCN (E). In contrast to the deletion of *Atoh1* (B, D, E) we have a different central projection that does not form a projection from the base and apex, but fibers reach out and overlap from all SGN fibers (F). Combined from (Filova et al., 2022a; Filova et al., 2020; Maricich et al., 2009; Muniak et al., 2016).

Hoxb2) results in *Hox2* genes are expressed in glutamatergic cells of the anteroventral cochlear nucleus without expression of cochlear hair cells (Karmakar et al., 2017). Deletion of *Hox2* results in a broad and reduced tonotopic precision that fails discrimination close to sound frequencies comparable to *Neurod1* or *Isl1* CKO deletions (Filova et al., 2022b; Macova et al., 2019).

In summary, SGNs project centrally in a tonotopic organization that can affect somewhat the precision in Ephrins, which show near normal *Egr2* and *Hoxb1* deletions but show a reduced central projection in *Npr2*. A profound deletion is the overlap of *Neurod1* and *Isl1* that have virtually eliminated a tonotopic organization (Fig. 37).

5.2. *Atoh1* defines the cochlea nuclei as well as electroreception and lateral line nuclei

Expression of *Atoh1* defines the auditory nuclei in all tetrapods and has a separate auditory system in teleosts that also have developed auditory nuclei. Evidence suggests that electroreception depends on *Atoh1* that also depends on lateral line nuclei (Fig. 32). Logically, one would like to see a simple replacement by electro- and lateral line nuclei by the auditory nuclei, an idea that was proposed before we understood *Atoh1* functions for cochlear/lateral line/electroreceptions. A possible relationship in tetrapods can be the following relationships:

1. Electroreception is the last developing central projection that develops after the lateral line nuclei develop (Figs. 9, 32). A few auditory fibers are incompletely segregated from vestibular nuclei in salamanders and caecilians. However, many salamanders and caecilians have never developed electroreception and lateral line nuclei (Fig. 19) that have also reduced the auditory fibers, remaining only the vestibular projections in derived caecilians and derived salamanders.
2. Frogs lack electroreception that develops lateral line nuclei (Fig. 19) that persists in certain frogs (*Xenopus*). In addition, later stages show auditory nuclei that reach the basilar papilla and amphibian papilla fibers which migrate away from unknown sources to receive auditory nuclei input. There is a possibility from previous electroreception nuclei could transform into auditory nuclei.
3. Auditory nuclei are known for amniotes that receive input from the basilar papilla/cochlea fibers to generate a tonotopic organization (Figs. 32, 36) that develops without lateral line and electroreception nuclei.
4. A simple replacement from electroreception/lateral line nuclei by auditory nuclei is unlikely for amniotes. Instead, it appears that the most likely explanation is using the auditory nuclei *de novo* in amniotes

(Figs. 32, 38) after the development of hair cells and associated neurons that provide the input to the *Atoh1* dependent cochlear nuclei.

How can we reconstruct the loss of electroreception/lateral line that derives from the same area that gives rise to the cochlea? Obviously, the cochlear nuclei in mammals (r2-5) are identical to the electroreception (r2-r5), suggesting a close longitudinal expression in the cochlea and the electroreception in mammals and salamanders/gymnophionans (Fig. 32). Most likely we can look at the novel generation in tetrapods that may have evolved separately consisting with the parallel evolution of the tympanic membrane (Carr, 2020).

The best description of the cochlear nuclei was provided by Matesz (Matesz, 1979). The details suggest a major central projection that starts at r2 (trigeminal) and ends at r5/6 with the glossopharyngeal auditory fibers. In addition, an ovoid area exists that begins on the r4 (VIIIth nerve) that extends past r8 (XII neurons). The cochlear nuclei are formed into two parts that receive all cochlear fibers, mainly from the basilar papilla, amphibian papilla, and lagena which also have a separate central projection referred to as saccule projections. Details on the vestibular central projections are provided in the vestibular fibers that has a limited saccule but a large lagena projection (Birinyi et al., 2001). While the earliest central projections have been described in higher order connections for hearing (Kelley et al., 2020; Simmons, 2019), no detailed analysis for the frogs that should be traced from pre-metamorphic to postmetamorphic

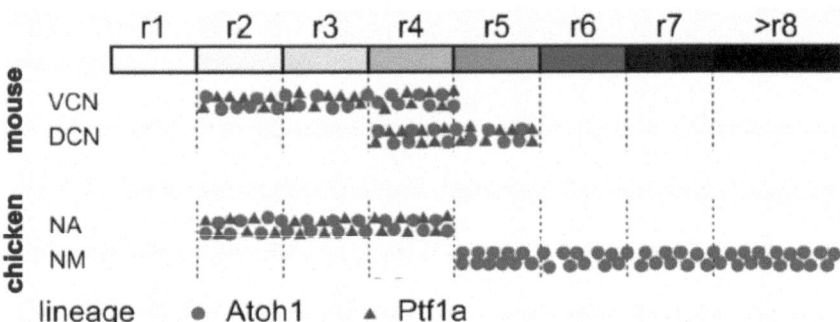

Figure 38: Mouse (top) and chicken (bottom) depend on *Atoh1* formation which has a different rostro-caudal extension. Both mice and chicken start at the ventral cochlear (ventral cochlear nucleus, VCN; nucleus angularis, NA) followed by the dorsal cochlear nucleus (DCN) that forms from r5 and some components from r4. Direct input for the nucleus magnocellularis (NM) is a caudal formation in chicken that is not found in mammals, that expands to r7/8. Note that a mix of *Atoh1* and *Ptf1a* in mice but have a simple origin from the NM. *Ptf1a* have a differential distribution according to Fujiyama et al. (2009). Modified after (Lipovsek and Wingate, 2018).

froglets. What is known is that most frogs lose the early development of hair cells and lateral line nuclei followed by the loss of sensory neurons (Wahnschaffe et al., 1987) whereas certain aquatic frogs (*Xenopus*) retain their lateral line next to the input of auditory projections. These data suggest that lateral line nuclei degenerate whereas frogs develop auditory nuclei independently. Salamanders and gymnophionans differ with respect to the loss of metamorphosis that is dominant in frogs. Lateral line and electroreception are present in all larvae in salamanders and gymnophionans that retain the sensor system during aestivation. Certain gymnophionans may have just the presence of permanent aquatic Typhlonectids whereas direct developing gymnophionans (caecilians) never develop lateral line and electroreception that is also correlating with a reduction of the auditory system due to the loss of a basilar papilla (Fritzsch and Wake, 1988). However, a single salamander is known to lose lateral line and electroreception after metamorphosis (Fritzsch, 1990) and develops an auditory input (Manteuffel and Naujoks-Manteuffel, 1990). In summary, the rostrocaudal extension requires proper tracing in metamorphic frogs to correlate the loss of lateral line and the formation of auditory nuclei. Most likely, it will have auditory nuclei between r2 to r8, more comparable to sauropsids (Carr, 2020).

Comparing the rostro-caudal extent overlaps between the cochlear nuclei in chicken (r2-r8) mostly with the lateral line projection (r1-r7/8; Fig. 32). Of course, neither the electroreception nor the lateral line showed no topological input in amphibians (Fig. 32) whereas mammals and chicken show a topological auditory central projection. We know that the cochlea ends up in the central projection in the cochlear nuclei in mice and chicken (r2) while the labeling of the caudal extension shows a further projection of the cochlear nuclei to reach beyond r8 in chicken (Fig. 38).

From here on, I will focus on the molecular basis of the mouse for which we have the most information available. We know that the earliest cochlear nuclei proliferate between ~E10-14. *Atoh1* is needed for cochlear nuclei formation that extends from the cerebellum to the spinal cord (Fig. 39). r2-5 generates all cochlear neurons also having additional neurons that migrate from the *Ptf1a* [Fig. 38 (Fujiyama et al., 2009)]. In addition, *Atoh1* is needed to generate the granule cells of the cerebellum from r1, collaborating with *Olig3*, among other genes (Lowenstein et al., 2022). Also, *Atoh1* is needed for generating rostrally from r6/7 to generate the pontine gray nuclei that project to the cerebellum while the inferior olive (r7-8), cuneate and gracile (r9-11; ION) reach out to end up to the cerebellum or generate the lateral lemniscus to project to the diencephalon (Diek et al., 2022).

A detailed analysis of the cochlear nuclei shows a more complicated distribution of r2-5 origin (Fig. 40). All *Atoh1* cochlear nuclei are dependent on dA1, developing as glutamate neurons (Glu). In addition,

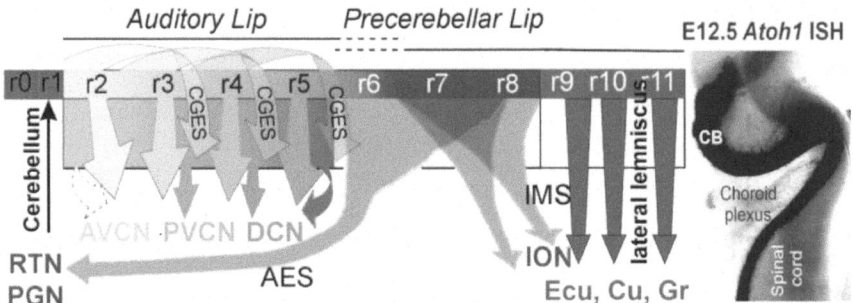

Figure 39: The organization of the primary central fibers forms from *Atoh1* which has a continuous central projection in the brainstem and spinal cord (right). *Atoh1* (left) shows the formation of the granular neurons (CB) of the cerebellum (r0-1), cochlear (r2-5, AVCN, PVCN, DCN), precerebellum (r6-8, PGN, RTN) and reaches out the lateral lemniscus fibers (r7-11) to become the gracile (Gr), cuneate (Cu) and the external cuneate (Ecu). The input from the inferior olive neurons (ION) projects to the contralateral cerebellum. Abbreviations: AES, anterior extramural stream; CGES, cochlear granule cell extramural stream; Cu, cuneate; Ecu, external cuneate; Gr, gracile; IMS, intramural migratory stream; ION, inferior olivary nucleus; ISH, in situ hybridization; PES, posterior extramural stream; PGN, pontine gray nucleus; RTN, reticulotegmental nucleus. Modified after (Fritzsch et al., 2022; Fritzsch et al., 2006).

we know a second population that derives from dA4/dB1 to migrate with dA1 to intermingle with the *Atoh1* derived neurons that depend on *Ptf1a* (Fujiyama et al., 2009). *Ptf1a* added two populations that are either expressed glycine (Gly) or are GABA (Fig. 40). Moreover, a distribution is provided to a certain degree with cochlear nuclei that depend on r2 (large and small spherical bushy cells that form a large calyx). The likely r3 dependent populations are globular bushy cells, and D- and T-stellate cells. In contrast to spherical bushy cells, globular bushy cells are interacting with large gaps in the calyx (Di Bonito and Studer, 2017; Oertel and Cao, 2020). r4 populations provide Octopus cells, many granule cells and Giant cells derived from dA1 that depend on *Atoh1*. r5 will depend on the formation of unipolar brush cells, fusiform cells, and granule cells that derive from *Atoh1* but has a large additional *Ptf1a* derived neurons that provide Golgi cells and ML-stellate cells (GABA) and have also provided the cartwheel and tuberculoventral cells that are glycine positive. Most importantly, the *Ptf1a* is highest number of r5 derived neurons whereas the r2-4 have a limited expression of *Ptf1a* derived neurons (Fujiyama et al., 2009). A description of the origin of *Ptf1a* shows a change of the dA4/dB1 that will upregulate *Lmx1b* to change the expression: instead of *Lhx1/5* and *Pax2* for dB1 it flips to *Lmx1b* expression. In addition, the dA4 positive *Lhx1/5* expands into a dA3 positive expression that expresses an enlarged *Lmx1b* and *Phox2b*. Moreover, the conversion shows a graded effect that

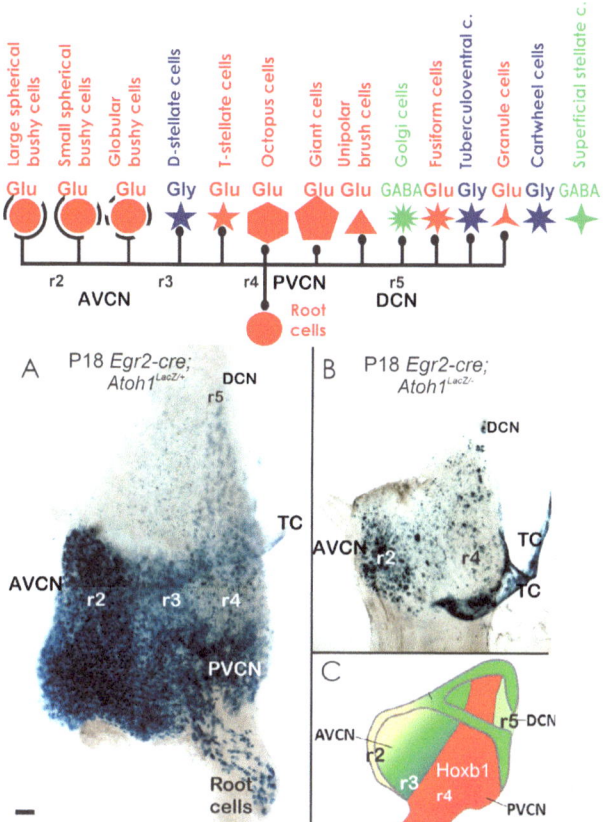

Figure 40: The cochlear nuclei depend on *Atoh1* neurons and develop in r2-5 that gives rise to small and large spherical bushy cells, globular bushy cells, T-stellate cells, octopus cells, giant cells, unipolar brush cells, fusiform cells, and granule cells (red). In addition, we have two populations that migrate dorsally from dA4/dB1 to provide ML stellate cells and Golgi cells (GABAergic; green) and cartwheel, tuberculate-ventral cells, and D-stellate cells (glycine positive cells; blue). A unique migration of the root cells that are positive for glutamate, like all other *Atoh1* derived neurons (top). The end bulb of Held has been indicated in the three cells of r2/3 bushy cells. Type I fibers connect all neurons while type II fibers reach selectively to granule cells (not shown). Comparing the $Atoh1^{LacZ}$ in the control (A) and using *Egr2-cre* show a selective loss of r3 and r5 (B) that conveniently show a different level of *Atoh1* level (A). No root cells developed and reduced the granule cells nearly entirely. Note the different insertion of the tela choroidea (TC), connecting with the choroid plexus, had a different insertion. r4 is highly positive and is likely present in the octopus and giant cells. *Hoxb1* expressed in r4 (C, red) extends from the root nearly to the top of DCN (compare A, C). In addition, the distribution is broad by granule cells. Modified after (Di Bonito and Studer, 2017; Fujiyama et al., 2009; Maricich et al., 2009; Oertel and Cao, 2020) and unpublished data.

converts r2-3 into *Lmx1b* positive neurons instead of dB1 whereas r4-6 is also converted from dA4 into dA3. Unfortunately, the later migration requires a detailed analysis of the cochlear nuclei (Iskusnykh et al., 2016; Elliott et al., 2023).

A paper detailing the loss of *Atoh1* after the expression of *Egr2-cre* or *Hoxb1-cre* (Maricich et al., 2009). *Atoh1* was shown using *Atoh1$^{LacZ/+}$* mice that were combined with the loss of *Atoh1* crossing them with floxed mice (*Atoh1$^{LacZ/flox}$*) to generate *Atoh1CKO* mice (*Egr2-cre; Atoh1$^{LacZ/flox}$* and *Hoxb1-cre; Atoh1$^{LacZ/flox}$*). Rhombomere 3 and 5 are lost in the *Atoh1* using the *Egr2-cre; Atoh1$^{LacZ/flox}$* (Fig. 40). There is a formation of the most anterior part of the AVCN that shows the highest level of *Atoh1^{LacZ}*. Conveniently, the level of r3 is less positive for control mice using *Atoh1^{LacZ}* (Fig. 40). Cells that are positive for *Ptf1a* are near normally developed after the deletion of *Atoh1* using either *Egr2* or *Hoxb1-cre*, consistent with the origin of dA4/dB1 derived neurons like the cartwheel neurons (Fujiyama et al., 2009). Some globular bushy cells develop that seems to be incompletely deleted after r3 deletion. Obviously, the octopus cells derived from r4 are present despite the loss of nearly all r5 derived neurons deleted (Fig. 40). In addition, we also have a deletion of *bHLHb5* to delete two cochlear neurons, unipolar brush cells and the cartwheel cells (Cai et al., 2016). Moreover, the ladybird homeobox protein homology (*Lbx1*) is expressed in certain cells, like superficial stellate cells and the cartwheel cells (Schinzel et al., 2021) consistent with the conditional deletion of *Atoh1* that show the formation of cartwheel cells (Maricich et al., 2009).

In summary, two populations generate from dA1 (*Atoh1*) and dA4/dB1 (*Ptf1a*) provide all neurons of the cochlear nuclei. Based on function can be positive for glutamate (*Atoh1*), glycine (*Ptf1a*, possibly from dA4), and GABAergic (*Ptf1a*, possibly from dB1). All neurons receive all *Atoh1* derived neurons that connect in various ways as end bulbs or terminating small connections. Moreover, type I and type II are parallel but only reach the granule cells from type II fibers that are somewhat like granule cells in the cerebellum.

5.3. Cochlea nuclei are unique in mammals

Neurons of the cochlear nuclei provide the first central input of the SGN fibers to begin auditory information. An overall similarity is shown in birds and mammals that have more different neuronal cell types compared to birds (Carr, 2020; Ryugo and Parks, 2003). A large set of cochlear neurons was described in mammals that reach out to interact with the endbulb of Held and other terminal interactions. In addition, cochlear nuclei provide information to the superior olivary complex (SOC), and many cochlear nuclei reach out to the lateral lemniscus and the inferior colliculi. In addition, *Ptf1a* derived neurons migrate into the cochlear neurons to provide inhibitory interactions using glycine and GABA mostly in r4 and r5 (Elliott et al., 2023).

The AVCN/PVCN input conveys information from an auditory input to eventually reach out to the SOC bilaterally and the contralateral lateral lemniscus. In contrast, the DCN acts as three layers to provide multisensory integration and reaches out directly to the contralateral inferior colliculus. Here I will describe the cochlear nuclei that are likely generated in r2-5 derived neurons, starting from bushy cells (r2-3; AVCN) to the T-stellate cells and the octopus (r3-4; PVCN), and will extend to the unipolar brush cells and fusiform cells (r5; DCN).

5.3.1. Bushy cells (AVCN)

Most of the rostral neurons are derived from r2 and found in large spherical bushy cells followed by the small spherical bushy cells that are positive for glutamate. The most popular globular bushy cells are spread between r2-r4 (Malmierca, 2015; Oertel and Cao, 2020). These three types of bushy cells form locally in the *Atoh1* positive dA1 neurons and differ in the endbulbs of Held that form a large calyceal area in the two spherical bushy cells but shows large gaps in the globular bushy cells that interact about 5 cochlear inputs from SGN. Globular bushy cells form gap junctions between neurons. In addition to receiving direct input from the cochlear SGNs by N-methyl-d-aspartate (NMDA) subtype of the glutamate terminals. In addition, bushy cells are also innervated by glycinergic and GABAergic terminals (Rubio, 2020) that receive inhibitory inputs from nearby local neurons and DCN (glycine, GABA), and also receive excitatory cholinergic fibers and non-auditory inputs (trigeminal input). Bushy cells are sharply tuned and respond to sound stimulation that enhances the encoding of the auditor SGN input and reflects the integration of its multiple auditory inputs. Overall, bushy cells enable these neurons to provide information on the timing and correlate the inputs to allow comparisons for interaural timing and intensity (Oertel and Cao, 2020).

Three kinds of bushy cells have been shown as the output to the SOC. Globular bushy cells project mostly to the contralateral medial nuclear trapezoid body (MNTB) but also provide the ipsilateral input to the lateral nuclear trapezoid body (LNTB). Large spherical bushy cells provide input bilaterally to the ipsilateral and contralateral medial superior olive (MSO). In contrast to bilateral outputs, the small spherical bushy cells provide input only from the ipsilateral lateral superior olive (LSO). I will deal later with the connection for the superior olivary complex (SOC) that interacts for information processing.

5.3.2. T-stellate cells (PVCN)

T-stellate cells are glutamate cells that develop from *Atoh1* and belong to the PVCN and are of r4 origin. T-stellate cells are organized in parallel to the auditory fibers along the isofrequency where they overlap with the bushy cells in their distribution. Collaterals of axons T-stellate cells are more widely dispersed compared to bushy cells and interconnected locally between AVCN and the DCN. T-stellate cells that are sharply tuned and fire tonically, and encode the amplitude of a sound. They are glutamatergic but also interact with local inhibitory D-stellate cells that also receive tuberculoventral cells from the DCN which are glycine (Oertel and Cao, 2020; Rubio, 2018a). Overall, the T-stellate cells play the role of spectral information for orienting to sound.

T-stellate cell axons reach out to the ipsilateral lateral superior olive (LSO), the contralateral ventral nucleus trapezoid body (VNTB), the contralateral ventral nucleus of the lateral lemniscus (VNLL), the contralateral inferior colliculus (IC). Minor branches can occur to reach the ipsilateral intermediate and dorsal lateral lemniscus (INLL, DNLL).

5.3.3. D- and L-stellate cells (PVCN)

D-stellate cells are distributed in the AVCN and reach out to the PVCN. They are inhibitory neurons that are glycinergic that migrate from *Ptf1a* from dorsal origins (dA4/dB1). Axons form local connections and span along the tonotopic axis. D-stellate cells affect the sound by interconnecting the ipsilateral with the contralateral cochlear nuclei on the two sides. Dye injections document that they reach out to the contralateral spherical and globular bushy cells. In addition, octopus cells and fusiform cells are targets for D-stellate cells that reach out to both ipsilateral and contralateral innervation (Oertel and Cao, 2020).

In addition, a large population of small glycinergic neurons are recently identified and form the L-stellate cells (Ngodup et al., 2020). More detailed analysis and defining of the development require the projection pattern of L-stellate cells.

5.3.4. Octopus cells (PVCN)

Octopus cells derive from r4 that generate *Atoh1* glutamate positive neurons. Dendrites cross auditory nerve fibers and send out tonotopic input of wide frequency tuning. The function acts as coincidence detectors that interact with the frequency sweep directions. They provide temporal interactions between the harmonics of different sound sources (Lu et al., 2022). Octopus cells receive very few inhibitory connections that are driven by glutamate inputs that reach out to the dendrites. More work is needed to detail the different inputs and how they respond in more detail to the octopus's function and how it relates to its targets.

The main projection becomes a separate bundle, the intermediate acoustic stria, and reaches out to the superior paraolivary nucleus (SPON) bilaterally (Malmierca, 2015) to end up in the contralateral ventral nucleus of the lateral lemniscus (VNLL).

5.3.5. Root cells (PVCN)

Root cells can be traced from the origin of r3/4 to migrate into the auditory nerves (Fig. 40) generated from *Atoh1* which is glutamate (Maricich et al., 2009) in rodents. Input forming boutons from auditory nerve fibers are mostly from the base of the cochlea. In addition, inhibitory, cholinergic, and noradrenergic inputs have been documented (Oertel and Cao, 2020). Root neurons are tuned to high frequencies that mediate a startle reflex.

The very large axons reach out to provide input to the contralateral pontine nuclei and reach out to the medullary reticular formation, intercollicular tegmentum, superior colliculus, ipsilateral facial nucleus, and the lateral paragigantocellular nucleus (Malmierca, 2015).

5.3.6. Fusiform (pyramidal) cells (DCN)

Fusiform neurons (also known as pyramidal cells) are developing in r5 and require *Atoh1* for their differentiation of glutamate neurons (Fig. 40). Dendrites interact with molecular layers whereas a basal dendrite reaches deeper and provides auditory fiber input. Deep fusiform dendrites are parallel in their orientation and receive a tonotopic input (Malmierca, 2015) that is sharply tuned and forms an isofrequency lamina (Trussell and Oertel, 2018).

The fusiform axons cross to the contralateral fibers to reach the inferior colliculus (Malmierca, 2015) and extend for direct input of the medial geniculate body (MGB).

5.3.7. Giant cells (DCN)

Giant cells depend on r5 that are *Atoh1* and are positive for glutamate. These cells are not organized tonotopically, as they spreads through several layers (Trussell and Oertel, 2018). They function distinctly from fusiform neurons that respond more broadly tuned, having a different threshold and lack a temporal response. The output is, like fusiform neurons, to reach out the contralateral inferior colliculus.

5.3.8. Granule cells and unipolar brush cells (VCN, DCN)

Rhombic lips are generated from r2-r5 to generate the external granular layer which also produces unipolar brush cells (UBCs) and granule cells in chickens and mammals (Iulianella et al., 2019). Moreover, the rhombic lip is known to interact with the electrosensory input from mormyrid fishes that extends across the external granular layer (Dempsey et al., 2019) known as the external granular layer in most gnathostomes. The uvula and nodule of the cerebellum and the cochlear nuclei interacts directly with UBCs and granule cells as they function as intrinsic amplifiers on granule cells. Two types of UBCs are found: The *ON*-type expresses mGluR to activate current whereas the *OFF*-type that are positive for calretinin (Elliott et al., 2021b) and responds to glutamate causes hyperpolarization (Trussell and Oertel, 2018). The thickness of granule cells can vary by forming granule cells in rats and mice but can be thin in primates that nevertheless exist in all mammals.

Like UBCs, granule cells depend on *Atoh1* and are glutamate positive. They are overlying the VCN as a shell to end up in the DCN parallel fibers to generate the molecular layer. The input is coming from type I SGNS and, most importantly, the type II cochlear axons (Muniak et al., 2016). Parallel fibers extend superficially and interact with the Golgi cells. An interaction results in an *ON*-type of UBC that drives the granule cells to inhibit Golgi cells. The output of Golgi cells that inhibit the intrinsic mossy fibers will, likely, be *OFF*-type UBCs. Further analysis requires additional understanding of the interaction between UBCs and granule cells.

5.3.9. Golgi cells and superficial stellate cells (DCN)

These two types of GABAergic neurons are provided from the *Ptf1a* to distribute among the DCN to interact with the fusiform neurons. Golgi cells interact with granule cells that provide axons to contact with granule cells to form a functional unit (Trussell and Oertel, 2018). Golgi cells receive a direct auditory input that also reaches out to nearby granule and UBCs cells and provides inhibitory synapses for feedforward and feedback inhibition.

Superficial stellate cells provide glycine/GABAergic inputs that are located in the molecular layer. These cells provide contacts with cartwheel and fusiform cells that interact with tuberculoventral (Rubio and Juiz, 2004). They will interact with these two types of hair cells that are less known compared to Golgi cells. They provide electrical gap-junctions with each other and the fusiform cells that can provide auditory information via the fusiform input (Trussell and Oertel, 2018).

5.3.10. Tuberculoventral and cartwheel cells (DCN)

Both tuberculoventral and cartwheel cells are inhibitory and glycinergic that are derived from *Ptf1a* and migrate into the cochlear nuclei. Tuberculoventral neurons have direct input from the auditory input that does not receive the cartwheel auditor input (Rubio and Juiz, 2004). Connecting local interaction with the fusiform neurons also have input from outside D-stellate cells (glycinergic) and into the ML-stellate cells (GABAergic). Some terminals are positive for glycine and GABA to provide an adult cell and terminal ends (Malmierca, 2015). Cartwheel-cartwheel interactions are slow, whereas the cartwheel-fusiform network is sensitive to transient changes (Trussell and Oertel, 2018).

The tuberculoventral system interconnects between the VCN and the DCN and contains frequency specific and diffuse projections to contribute to encoding the sharp spectral sensitivity. Tuberculoventral cells are the lowest layer that is interacting with dendrites of fusiform neurons that receives auditory input as excitation but also have a direct auditory input through the T-stellate cells and receive inhibitory input from D-stellate cells that are inhibited by glycine. They encode sharp spectral sounds.

5.3.11. Multimodal output and inputs

The DCN receives multisensory inputs turning it into a multimodal processor. The primary input from type I and type II axons almost reach the cochlear nuclei, except for cartwheel cells and superficial stellar cells (Fig. 41). The granule cells also provide UBCs while UBCs may be activated or inhibit granule cells to give rise to the parallel fibers of granule cells. In addition to cochlear afferents, we know that vestibular and ganglia, dorsal column nuclei, dorsal root ganglia, trigeminal ganglia and nuclei, inferior colliculi, pontine nuclei, and auditory cortex fibers reach out to interact with nearly all cochlear nuclei (Oertel and Cao, 2020; Rubio, 2018a). The projection forms into three branches: the dorsal acoustic stria (fusiform cells, T-stellate cells), the intermediate acoustic stria (octopus), and the ventral acoustic stria (spherical bushy cells, globular bush cells, T-stellate cells, root cells). A unique output reaches the contralateral cochlear nucleus that inhibits contacted neurons, the D-stellate cells. Two projections reach out to the IC (fusiform cells, T-stellate cells). In addition, the octopus cells reach the contralateral ventral nucleus of the lateral lemniscus (cVNLL) which also receives branches from the T-stellate cells. Three major inputs are provided for the superior olivary system (SOC): lateral superior olive (ipsilateral LSO) and medial superior olive (ipsi- and contralateral MSO) provide inputs from the large spherical bushy cells while the small spherical bushy cells provide input to the ipsilateral lateral superior olive (iLSO). The medial nucleus of the trapezoid body (contralateral MNTB) is the main output of the globular bushy cells that also receive a minor ipsilateral lateral nucleus of the trapezoid body (iLNTB). T-stellate cells are the most output that reach the contralateral medical geniculate body (cMGB), the contralateral inferior colliculus (pIC), the contralateral ventral nucleus of the lateral lemniscus (cVNLL), the contralateral ventral nucleus

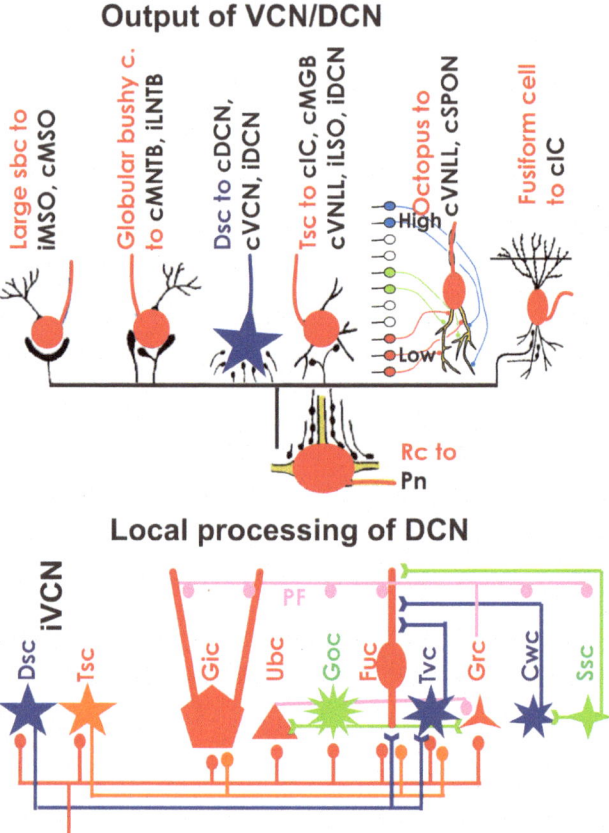

Figure 41: (Top) Type I and type II afferent fibers reach out to innervate branches of the cochlear nucleus in a tonotopic input. While type I differ in terms of the complexity of terminal endings while type II form simply boutons terminals only. A large set of direct input is provided directly from large spherical and globular bushy cells, T- and D-stellate cells, octopus (cSPON), and fusiform neurons (cIC) that reach out the different output to the superior olivary complex (SOC), lateral lemniscus (LL), inferior colliculi (IC) and pontine nucleus cells (PnC). (Bottom) The DCN is an integration of direct (type I + II) and indirect input that is in part excitatory (glutamate, shades of red) and part inhibitory interactions (glycine, blue; GABA, green). Only excitatory input provides nearly all DCN, except for the cartwheel and the superficial stellate cells. However, the granule cells provide parallel fibers to reach out to nearly all, except for the tuberculoventral cells and the UBCs. Golgi cells receive direct input from auditory fibers and receive input from granule cells that give rise as inhibitor neurons to modulate certain UBCs and certain granule cells. Note that certain electrical gap-junctions are known in Golgi, fusiform, and superficial stellate cells. For abbreviations see Fig. 40. Modified after (Lu et al., 2022; Malmierca, 2015; Oertel and Cao, 2020; Rubio, 2018a; Rubio and Juiz, 2004; Trussell and Oertel, 2018).

of the trapezoid body (cVNTB), the ipsilateral lateral superior olive (iLSO) and the fibers connect from VCN to reach out the DCN input, shared with D-stellate cells (Fig. 41). In addition, the root cells provide the outputs of a large fiber that reaches out directly to the contralateral reticular pontine nucleus (Pn). The newly described L-stellate cells interact with the ipsilateral VCN (Ngodup et al., 2022).

The main function of these cochlear nuclei provides acoustic information to reach out the separate, parallel, and ascending pathways. Root neurons mediate the startle reflex. Bushy cells sharpen the fine structure of sounds, differential between low and high frequency T-stellate cells form an envelope of sounds while octopuses respond to sweeps between different frequencies. Fusiform neurons integrate excitatory and inhibitory processing that receives input from deep layers and superficial parallel fibers.

Multimodal input signals range from the fusiform output and multimodal input that reach out from the DCN. Local connections are provided by almost all auditory nerve fibers while the T-stellate cells (excitatory) and receive D-stellate cells (inhibitory) extend to the DCN (Fig. 41). The complexity of integration can differ between neurons. Certain neurons do not have a direct auditory fiber (cartwheel cells, superficial stellate cells) that may have one (UBC, granule cells, Golgi cells), two (giant cells) or more direct or indirect auditory inputs (fusiform cells, tuberculoventral cells). Parallel fibers (PF) are interacting with nearly all input, except for UBCs and the tuberculoventral cells (Fig. 41). UBCs form ON and OFF cells that can activate or inhibit granule cells that also interact with the Golgi cells. Electric communication exists between superficial stellate cells and fusiform cells while the Golgi cells show interactions with each other.

In addition, glutamate receptors differ in different synapses that form at the position. For example, the fusiform cells receive parallel input near the apical dendrites and a basal direct auditory input from the SGNs. In addition, the input from the pinna muscles and trigeminal input provides a complex output. Moreover, innervation by dopaminergic, noradrenergic, and serotonin fibers provides a distinct action.

Despite for the importance for hearing, it is the most common hearing loss that affects about 1.5 billion people. We know that hearing loss starts in humans by first losing the outer hair cells in the base and eventually losing the inner hair cells as well. The loss of hair cells will eventually result in a loss of spiral ganglion neurons (Elliott et al., 2022a). Hearing loss can result in the ratio of excitation and inhibition that is more affected in the VCN in older mice (Rubio, 2018b) which requires a more complex analysis combining connectional, electrophysiological, and molecular information.

In summary, the VCN and DCN provide monaural sound localization, and integration of auditory signals, and provides various regulation of the

DCN circuitry. A primary tonotopic input provides principal cells that will fine tune by inhibitory neurons. Future work should address the tonotopic input for octopus cells that has a unique output and will provide through modern anatomical and physiological studies to help the VCN and DCN neuronal network.

5.4. Outflow of cochlear nuclei to SOC

A second station receives from the auditory SGNs that reach out from auditory nuclei to be innervated by the superior olivary complex (SOC). Of the five to nine nuclei belonging to the SOC, at least four nuclei are described in detail (Kandler et al., 2020): the medial nucleus of the trapezoid body (MNTB), the medial superior olive (MSO), the lateral superior olive (LSO) and the superior paraolivary nucleus (SPON).

Cells are generated from E10.5 to E14.5 in mice but have additional formation generated later and found at E17.5. An origin is produced in the cochlear nuclei from which most SOC are migrating to distribute among the SOC and also migrate to the ventral nucleus of the lateral lemniscus [VNLL (Di Bonito and Studer, 2017)]. The origins are, in part, dependent on *Atoh1* but also added to the homeobox engrailed-1 [*En1* (Milinkeviciute and Cramer, 2020)] to add some unknown origin that requires detailed analysis (Lipovsek and Wingate, 2018). Many of the SOC are generated mostly in r5 but also have a small population that adds from the LSO (Di Bonito and Studer, 2017). An incomplete deletion of r5 using *Krox20* (*Egr2*) deletion of *Atoh1* shows a small population remains in a reduced SOC input (Maricich et al., 2009). It is unclear how many SOC neurons are primarily lost or are a consequence of the loss of SGNs that were studied on P18 or older mice. Obviously, all cochlear nuclei disappear in an *Lmx1a/b* DKO mice that interact with the choroid plexus and the rhombic lip. Further investigation is needed to detail the formation of SOCs without *Lmx1a/b* DKO null mice (Chizhikov et al., 2021).

In summary, the SOC is in part derived from the cochlear nucleus anlage. A full description of all brainstem neuron development would be needed to fill the origin of SOCs. A unique connection is obvious from r4 that generates octopus cells that migrate to the VNLL.

5.4.1. Lateral superior olive (LSO)

Neurons are tonotopically arranged in an S-shaped row of LSO neurons and play a role in encoding interaural sound level differences (ILDs), mostly connecting with higher frequency. Binaural interactions enable LSO to process ILDs. Five to seven different cell types have been identified that receive mostly elongated neurons (Malmierca, 2015). In addition to receiving excitatory, ipsilateral, glutamatergic spherical bushy cells from the AVCN they also receive excitatory and inhibitory, contralateral, and ipsilateral dorsal nucleus of the lateral lemniscus (DNLL) and the inferior colliculus (IC). A major input is provided from the contralateral, globular bushy cells to reach the MNTB which inhibits the glycinergic neurons of LSO to allow the precise interaction from the excitatory postsynaptic potential (EPSPs) with IPSP to underlie the ILD. Inhibition of MNTB can completely block LSO excitatory input (Kandler et al., 2020) which requires precise, simultaneous temporal stimulation. Elicit of GABA and glutamate is released, in addition to glycine, that function is somewhat unclear, possibly acting on long-term depression and sharpening the tonotopic organization. The output innervates the dorsal nucleus of the lateral lemniscus (DNLL).

5.4.2. Medial nucleus of the trapezoid body (MNTB)

The input from globular bushy cells projects with large axosomatic, glutamate positive, and single calyces of Held end up in the contralateral MNTB neurons (Kandler et al., 2020). The size of the somatic synapse forms unique pre- and postsynaptic input that each contains hundreds of active zones. MNTB principal neurons generate a sustained pure tone response that shows phase-locking signals that show a topographic organization along the medio-lateral axis: low frequency is located laterally whereas high frequency is located medially. Low-frequency tones and amplitude provide temporally phase-lock inhibition while higher frequencies do not phase locking but become phasic-tonic onset responses. The major output provides inhibitory input to the ipsilateral LSO and MSO to extract interaural time and intensity differences. In addition, the MNTB neurons encode the temporal structure of sound (SPON). The collateral output of the MNTB adds extensive tonotopic damping by using inhibitory connections. Beyond the input-output of the principal MNTB neurons, a large ipsilateral LSO, MSO, SPON, VNTB, and LNTB are interconnected with the ventral nucleus of the lateral lemniscus [VNLL (Malmierca, 2015)].

5.4.3. Medial superior olive (MSO)

The MSO is located between the LSO (lateral) and the MNTB (medial) that aligned bipolar MSO primary neurons to integrate bilateral spherical bushy cells (AVCN) with bilateral globular bushy cell input, a four-way comparison: Bilateral excitation reaches out from the MSO while ipsilateral inhibition (LNTB) and contralateral inhibition (MNTB) interact with similar latencies in rodents (Fig. 42). MSO neurons provide the interaural time differences (ITDs) detection to localize sound sources in the azimuth (Kandler et al., 2020). It appears that different speed conductance may be the basis for the ITD permitting the excitatory and inhibitory input to allow correlation as coincidence detectors for the binaural input. The spatial coding of the sound of MSO connectivity combined with synaptic specialization for speed and timing renders these neuronal coincidence detectors for interaural phase differences (IPDs) for detection (Pecka and Encke, 2020). The output is mainly via the ipsilateral dorsal nucleus of the lateral lemniscus (DNLL) and the central nucleus of the inferior colliculus [IC (Malmierca, 2015)].

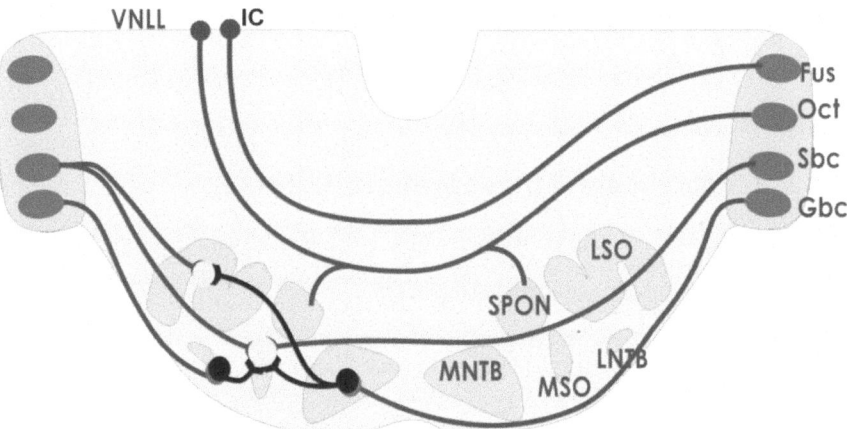

Figure 42: Major outputs of the AVCN/PVCN/DCN neurons reach the SOC complex, lateral lemniscus, and inferior colliculus. Details are the binaural sound that shows the LSO which has only a single ipsilateral spherical bushy cell (Sbc) and also receives a contralateral globular bushy cell (Gbc) that is inhibited from MNTB neurons. In contrast, the MSO receives bilateral ipsi- and contralateral excitatory input from the Sbcs and receives bilateral ipsi- and contralateral inhibitory input from Gbc's via the MNTB and the LNTB. In addition, the octopus cells (Oct) provide bilateral input to the SPON and continue to end up with VNLL. The fusiform output from the DCN bypasses the SOC to reach out to the contralateral IC. Not shown are the T-stellate cells that end in certain parts of the SOC, the VNLL, and the IC. Abbreviations: AVCN, anteroventral cochlear nucleus; DCN, dorsal cochlear nucleus; Fus, fusiform neurons; Gbc, globular bushy cells; IC, inferior colliculus; LSO, lateral superior olive; LNTB, lateral nucleus of the trapezoid body; MNTB, medial nucleus of the trapezoid body, MSO, medial superior olive; Oct, octopus cells; Sbc, spherical bushy cells; SPON, superior paraolivary nucleus; VNLL, the ventral nucleus of the lateral lemniscus. Compiled from (Kandler et al., 2020; Malmierca, 2015; Pecka and Encke, 2020).

5.4.4. Superior paraolivary nucleus (SPON)

The SPON in rodents is referred to as dorsomedial paraolivary nucleus (DMPO) which is also found in bats, cats, and primates but is best characterized in rodents. Multipolar GABAergic neurons and small glycinergic neurons are described. Contralateral excitatory input is provided from octopus neurons and a few T-stellate cells (Fig. 41) that are provided by the SPON (Kandler et al., 2020). In addition, the contralateral input provides the ipsilateral T-stellate cells and some globular bushy cells. Octopus are broadly tuned (Lu et al., 2022) that integrate inputs over a wide frequency range now known to affect frequency responses to neurons which provide precise timed phasic ON responses. Inhibition is provided by glycinergic, ipsilateral MNTB, and the LNTB and provides local GABAergic input. From SPON exits to innervate the ipsilateral IC and reach out the tectal longitudinal column, comparable to birds (Carr, 2020) and the lateral line input from *Xenopus* (Dean and Claas, 2020), that may provide a tonotopic output (Malmierca, 2015). The phasic response of neurons suggests they are tuned to sound intensity or for tracing the temporal envelope of sound that may be responsible for T-stellate neurons interfering with the precise timed OFF responses. SPON is to encode rhythmic sound patterns that provide temporal periodicity which may lead to specific vocalization and speech signals.

In addition, we need to mention the periolivary nuclei (PO), the ventral nucleus of the trapezoid body (VNTB), and the lateral nucleus of the trapezoid body (LNTB). For example, the PO receives VCN bilaterally from T-stellate and octopus vells, from the ipsilateral MNTB and ipsilateral IC. PO projects to VCN and DCN to provide reciprocal connections. Their connections are detailed elsewhere (Malmierca, 2015).

5.4.5. An integrated perspective of the SOC

The auditory system is optimized for coding temporal information within microseconds. In addition, we must provide for the temporal resolution (microseconds) and the temporal integration (milliseconds), and add the temporal abstraction in seconds (Kopp-Scheinpflug and Linden, 2020). Three auditory inputs drive sound processing: interaural sound level differences (ILD), interaural time differences (ITD) and interaural phase differences (IPDs (Pecka and Encke, 2020). Both ITD and ILD cues can be used for sound localization and both processing requires the comparison to extract specific sound properties that compare the left and the right ear using the auditory brainstem, the cochlear nuclei, and the SOC.

Binaural processing is needed for the precise interaction of glutamatergic excitations and compared with glycinergic inhibition. Comparison of timing of excitatory and inhibitory inputs plays a role in the LSO where it acts as an integral part of computation. Timing changes are generated by both ITD and ILD in the LSO and use binaural processing (Fig. 42) consisting of input strength (amplitude) and timing. LSO contribute to ITD coding.

The MSO has four direct input comparisons: two ipsi- and contralateral excitatory inputs from the spherical bush cells and has two ipsi- and contralateral inhibitory inputs via the globular bushy cells (Fig. 42) for synaptic transmission that facilitates microsecond ITD sensitivity. A relative increase in activity compared with the four channels may indicate a corresponding contralateral location from ITD sensitive channels. In addition, the MSO provides sensitivity to interaural phase differences (IPD). MSO responds to a particular pure tone frequency and provides an effective delay of the preferred IPD to isolate sounds which captures a mechanistic function. The pattern of spontaneous activity can influence the refinement of projections from MNTB to lateral superior olive (LSO). Recent evidence suggests the LSO and MSO can be used in the sound source azimuth, needs to coherent midbrain representation and are combined to yield sound source location (Pecka and Encke, 2020).

A large family of 12 genes (Kv1-Kv12) provide potassium channels in five classes. In addition, the HCN gene family encodes the hyperpolarization of cation channels (Kopp-Scheinpflug and Linden, 2020). Low-frequency of each cycle can follow the fire response to every cycle, referred to as phase-locking. Higher sound frequencies (above ~2 kHz) are no longer phase-locked responses resulting in peri stimulus time histograms (PSTHs). Action potential (AP) can skip individual nerve fibers but can result in a reliable response between several fibers to generate a population response of PSTH. In addition, frequency changes can affect both the temporal fine structure and the spectral content of acoustic signals in the brainstem. I will deal with auditory processing after the lateral lemniscus and the inferior colliculus have been defined.

5.5. Lateral lemnisci are generated from two distinct populations

Three layers of nuclei are described: the ventral nucleus of the lateral lemniscus (VNLL), the intermediate nucleus of the lateral lemniscus (INLL), and the dorsal nucleus of the lateral lemniscus (DNLL). I will follow the presentation of Felmy and Meyer (Felmy and Meyer, 2020) which uses a different nomenclature (Malmierca, 2015).

Development is following in two distinct neuron populations. VNLL comes from r4 and migrates rostrally, providing a unique anterior migration (Di Bonito and Studer, 2017; Di Bonito et al., 2017). Most neurons are either glycinergic or GABAergic that likely derive from *Ptf1a*, consistent with the migration of inhibitory inputs to the cochlear nuclei and the SOC (Fujiyama et al., 2009; Lipovsek and Wingate, 2018). In addition, the INLL is glutamate, excitatory neurons whereas the DNLL is mostly derived from GABAergic neurons (~80% of neurons). The origin of these neurons is unclear (Di Bonito et al., 2017).

5.5.1. Ventral nucleus of the lateral lemniscus (VNLL)

The VNLL output comes from the octopus (PVCN) and the T-stellate cells (AVCN, PVCN), and bushy cells (AVCN). In addition, fibers reach out from the MNTB and, lesser, the LNTB and the SPON, some of which resemble small calyces of Held, mostly from octopus cells (Fig. 43). The output will extend the IC and has collaterals to INLL and DNLL and provides local interactions of VNLL (Malmierca, 2015). Fibers from the VNLL reach out to IC bilaterally and extends to the medial geniculate body (MGB) directly and continue further to the optical tract (Di Bonito et al., 2017). The tonotopic organization is disputed but it is documented for bats that receive directly globular bushy cells that provide a tonotopic gradient (Felmy and Meyer, 2020).

VNLL responds rapidly and with high precision to sounds that are modulated. Excitatory glutamate provides the input for the VNLL while inhibition is mediated by GABA and glycine receptors. VNLL acts as monaural input with little binaural processing. Octopus cells likely provide their frequency tuning (Lu et al., 2022). Octopus-VNLL circuit is crucial for speech processing (Oertel and Cao, 2020). Feedbacks provide by IC to modify temporal precision where it is broadly tuned its signal. Overall, the VNLL detects temporal patterns to generate feed forward inhibition to modify the sound evoked excitation.

5.5.2. Intermediate nucleus of the lateral lemniscus (INLL)

Wedged in between the VNLL and the DNLL, the INLL is best characterized by receiving excitatory, glutamatergic neurons and lacks GABA and glycine input (Felmy and Meyer, 2020). INLL receives from globular bushy cells and T-stellate cells from the VCN, which is tonotopically organized in bats, and receive also from the superior olivary complex (Fig. 43). A major input comes from the IC and also receives collaterals from VNLL. Information between different frequency channels might affect acoustic feature processing by INLL neurons.

5.5.3. Dorsal nucleus of the lateral lemniscus (DNLL)

Immediately adjacent to the IC is the DNLL which intersects with the commissure of Probst bilaterally (Fig. 43). ~80% of DNLL neurons are GABAergic while ~20% are unclear in their transmitter type. The DNLL is tonotopically organized and processes frequencies differentially in some but not all mammals (Malmierca, 2015). Both cochlear nuclei and the SOC, including SPON, are reaching out to the DNLL while it also receives collaterals from VNLL and INLL. Descending tonotopic information reaches out from the auditory cortex and the IC, possibly acting as top-down

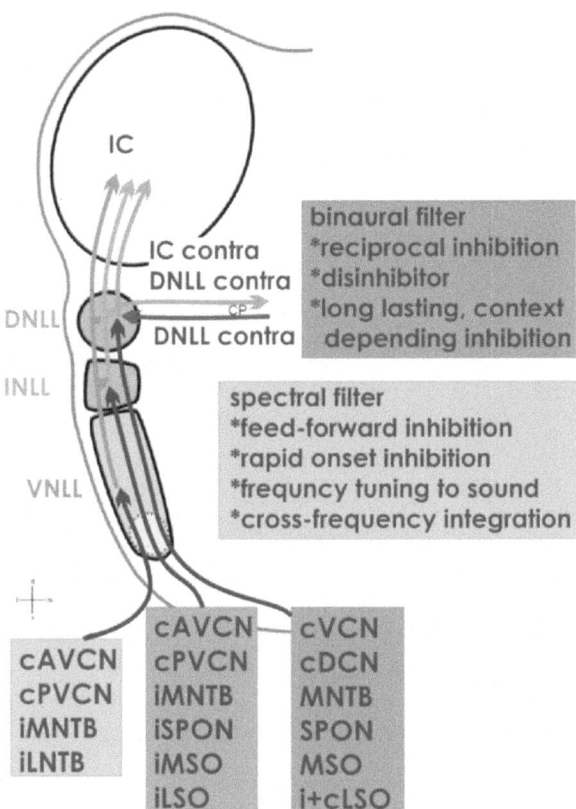

Figure 43: The lateral lemniscus provides three subnuclei. VNLL receives contralateral AVCN/PVCN and also receives input from MNTB and LNTB which receives large glutamatergic neurons (dotted circle) that mostly reach out from octopus neurons. Temporally precision will provide a feed forward to drive the spectral filter in the VNLL. INLL receives from VCN and all SOC projections and is involved in cross-frequency integration. Like the VNLL, collateral provides also the DNLL. The DNLL obtains input from VCN and DCN, the SOC, and provides collaterals from VNLL and INLL. A bilateral input-output is provided from the commissure of Probst (CP) that also sends fibers to the IC which provides feedback from the IC. The DNLL acts as a binaural filter in the suppression of sound source localization. C, contralateral; I, ipsilateral. Modified after (Felmy and Meyer, 2020).

control (Felmy and Meyer, 2020). DNLL provides excitatory glutamatergic and receives inhibitory GABA and glycine input. GABAergic inhibition from the contralateral DNLL interacts with glutamatergic excitation and is the basis of its circuit function and binaural filter (Fig. 43).

Tuning of DNLL is provided by ascending inputs from SOC. ITD and ILD coding from SOC reflect binaural interactions via the commissure of

Probst. Long-lasting inhibition between the DNLL hemispheres and the IC can lead to reciprocal inhibition that is crucial for binaural hearing (Felmy and Meyer, 2020). Ipsilateral sound stimulation from the contralateral DNLL suppresses the ipsilateral DNLL (Pecka and Encke, 2020). Moreover, successive sounds could be detected faster, enabling faster reaction time.

Since the lateral lemniscus is integrating with the inferior colliculus, I will deal with the integration after the role of IC is described.

5.6. The inferior colliculus (IC)

Immediately caudal of the superior colliculus (SC) of the midbrain and rostral to the cerebellum is the inferior colliculus (IC). The IC receives from almost all cochlear nuclei, the SOC and the LL. The output is the almost sole source of the medial geniculate body (MGB), which reaches out to the auditory cortex. A commissural interconnection receives non-auditory input. The IC is an integrative center for auditory input that can be demonstrated for selective lesions of the auditory spiral ganglion neurons (Filova et al., 2022b). The IC can define separate subnuclei: the ovoid central nucleus of the IC (ICc) has a dorso-medial cortex (ICd), a ventro-lateral cortex (ICl), and has the most rostral cortex [ICr (Malmierca, 2015; Rees, 2020)].

The IC develops as a subdivision of the SC/IC and is identified as a dorsal part of the midbrain [m1 (Nieuwenhuys and Puelles, 2015)]. Expression of genes plays a role in IC development such as the open brain (*Opb*) or *Shh* deletions (Eggenschwiler et al., 2001). Most important are *Otx2, Lmx1b, Wnt1,* and *Engrailed 1+2 (En1,2)* that are defining the development of the IC and SC (Glover et al., 2018). The complete loss of the *Wnt1* deletion, that is downstream of *Lmx1b*, is eliminating the entire midbrain (Mastick et al., 1996). Among several gene expression are *Fgf8 and Pax*2, among others, that reduce the IC (Driscoll and Tadi, 2022; Urbánek et al., 1997; Watson et al., 2017). Conversely, overexpression of *Shh* using *Smoothened* (*Smo*) results in an exuberant development of the IC and the cerebellum (Jahan et al., 2021). Recently it was generated from *Isl1* using *Pax2* drivers that cause a reduced size of the IC and resulted in incomplete disruption of normal function (Chumak et al., 2021). Downstream of *Pax2* are *Pax3, Pax7*, and *Meis2* which interact with the IC roof plate development [Fig. 44 (Agoston et al., 2012; Nakamura, 2020)]. *Otx2* is driven by the expression of bHLH genes. Among the genes that are required are *Sox2, Ascl1, Neurod1, Neurog2,* and *Dll3* that interact with early initiation and differentiation of the IC (Henke et al., 2009; Kim et al., 2011). *Dbx1* deletion reduces the IC dramatically (Tran et al., 2023).

Downstream of *Neurog1* and *Neurog2, Neurod1* is expressed in the IC and can be used to identify the brachium of the inferior colliculus [BIC (Gurung and Fritzsch, 2004)]. Beyond that early description, we have

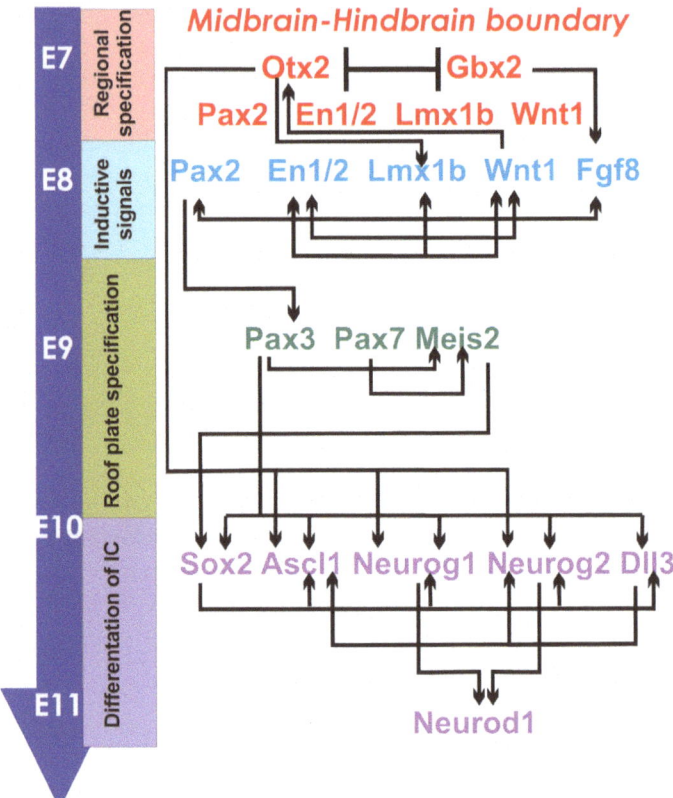

Figure 44: A genetic network controls the midbrain that defines through *Otx2-Gbx2* negatively interaction. Among downstream dependent genes is *Lmx1b* and bHLH gene regulation. Downstream regional specifications are inductive signals to pass these genes. Among them are genes needed for the midbrain development that regulate their expression. For example, *Wnt1* deletion results in midbrain deletion. Downstream is the interaction for *Pax3* and *Meis2* that are needed for normal development of the IC. A large set of bHLH genes that interact with *Dll3* to regulate as feedforward and feedback the respective gene level of expression. Downstream is *Neurod1* that requires detail deletion. Image was modified after (Agoston et al., 2012; Glover et al., 2018; Henke et al., 2009; Kim et al., 2011; Urbánek et al., 1997).

downstream glutamate, glycine and GABAergic expression in the IC that also depend on the neurotrophins for neuronal survival.

5.6.1. The central nucleus of the inferior colliculus (ICc)

The ICc provides its tonotopic organization. Ascending afferents originate from ventral and dorsal nuclei (VCN, DCN), and provide fibers from the superior olivary complex (SOC) and the lateral lemniscus (LL). The IC is monaural input from the cochlear nuclei and the VNLL whereas LSO and

DNLL are binaural (Fig. 45). T-stellate cells send from VCN bilateral, low frequency regions while high frequency receives only from the contralateral side. Likewise, the DCN projects bilateral input from fusiform, giant, and T-stellate cells. During development, afferents reach out prior to the onset of function and set up the input from ICc which is functional and ends up in segregated inputs. Detailed analysis shows anterograde of MSO and LSO receives mostly contralateral input. In the caudal region, the MSO terminal overlaps with ipsilateral LSO fibers while showing colocalization of rostral lamina co-occurred between ipsilateral and contralateral LSO. Excitatory, glutamatergic may form clusters with respect to their input patterns while inhibitory neurons are more homogenous (Rees, 2020).

At least two principal types of neurons have been identified (Malmierca, 2015) that have no clear correlation between the morphology of neurons and the physiology of cells. The organization is flattened neurons along the frequency bands that are about 200 um thick. Frequency presentation can be best demonstrated by electrode penetration that shows the tonotopic frequency from distinct sound input that can demonstrate the best frequency input (Fig. 45). Overall, the frequency band is distinct in lamina distribution that fits well both anatomical and physiological information (Rees, 2020) that are highly organized and integrate both spectral and temporal signals. Frequency tonotopic organization is affected by reduced input from the SGNs that result in the near-existence of a tonotopic input/output relationship (Filova et al., 2022b; Macova et al., 2019).

5.6.2. The lateral nucleus of the inferior colliculus (ICl)

Lateral nuclei are marked by a transition between nicotinamide adenine dinucleotide phosphate diaphorase (NADPH) in the superficial layer and have a cytochrome oxide (CO) rich centre of the ICl. Three layers can be distinguished that blend between the ICr and the ICc. The organization of dendrites differs from ICc in their thickness and orientation. Auditory fibers reach out from the cochlear nuclei and have feedback from the auditory cortex. The ICl receives also MGB input. Besides ipsilateral auditory input are fibers from the cuneate, the trigeminal nuclei, substantia nigra, parabrachial region, the midbrain central gray, the periventricular nucleus, the globus pallidus and the locus coeruleus (Malmierca, 2015) that provides a multitude of input.

5.6.3. The rostral nucleus of the inferior colliculus (ICr) and dorsal nucleus of the IC (ICd)

The rostral part of the IC has no distinct dendritic orientation that are mostly multipolar cells. The border between ICc and ICr is defined as a rostral extension and differs from ICl by unpolarized dendritic orientations.

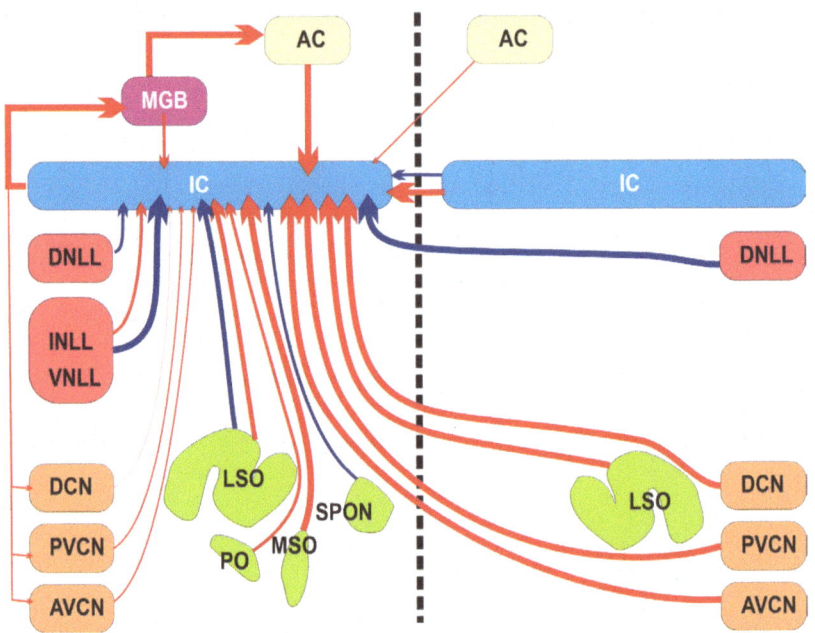

Figure 45: Major input to the IC and certain aspects of feedback. Note that the MGB is receiving the input from the IC which, in turn, projects to the AC. Red fibers are glutamatergic, blue is glycinergic/GABAergic. The thickness of fibers is proportional to their strength. Note the dotted line indicates the two sides of the brain. Modified after (Malmierca, 2015; Rees, 2020).

The ICd is found between the dorsomedial and caudal aspects of the ICc. Like with ICl, three layers can be distinguished. Like ICl, ICd is expressed nitric oxide that is NADPH staining that forms a gradient of distribution running from dorsal to ventral of ICc. The auditory cortex generates a tonotopic banded pattern, comparable to ICc. Tonotopically isofrequency contours continue and overlap with the ICc.

5.6.4. Commissural interconnection and neurochemistry

The IC receives interconnections between the two hemispheres defining commissural fibers that are intrinsic connections. Tracing the connections shows they are parallel to the isofrequency contours of ICc that will extend to the ICd. A small injection of tracers shows a mirror sheet from the contralateral side, arguing that the connection is homotopic isofrequency lamina of both sides of the IC. The largest fibers of commissural connections are glutamatergic neurons and have a small proportion of neurons that are GABAergic.

In the rat, the IC has about 25% of neurons which are GABAergic and contains glycinergic neurons, the rest are excitatory glutamate neurons that provide both AMPA and NMDA receptors. In addition, serotonin and noradrenergic terminals are present that comes from locus coeruleus and the dorsal raphe nucleus.

The spatial location of sounds is not topographically mapped, like the retinal mapping, but requires computations within the auditory brainstem and midbrain to extract cues involved in spatial analysis. IC receives input from MSO and LSO to extract binaural time and level differences for localization in the azimuth plane that also receives DCN input to provide spectral cues to detect sound elevation (Rees, 2020). Interaural time difference (ITD) provides the difference in time of arrival of sound between the ear and interaural phase difference (IPD) can be used to detect the direction of sound. Interaural level differences (ILD) are provided by the balance of binaural inputs between the ear, excitatory output from the contralateral IC while inhibitory output to the IC. ILD functions depend on the binaural level that may provide by IC comparison to help to overcome disambiguate. However, ILDs involve more complex interactions. Responses to amplitude modulation (AM) and frequency modulation (FM) may synchronize firing that shapes modulation transfer functions (MTF). MTF may help with directions in bats but does not show directional selectivity in mice and rats (Rees, 2020). The shape of the MTF will depend on background noise that also influences the best frequency stimuli.

Both the auditory cortex (AC) and medial geniculate body (MGB) send descending projections to the IC that is more prominent from the AC (Fig. 45). AC are predominantly ipsilateral that are glutamatergic and a smaller portion of GABAergic input to the ICc that could deliver IPSP from GABAergic neurons while glutamatergic EPSPs fire simultaneously (Ito and Malmierca, 2018; Malmierca, 2015).

Electrical stimulation of cortical neurons provides enhanced responses of IC neurons. The stimulus-specific adaptation follows a repeated stimulus which suggests an emergency property of the auditory cortex, some of which do not derive from the auditory cortex. It has been suggested that stimulus-specific adaptation (SSA) may be a component of short-term memory (Malmierca, 2015).

In summary, the interaural time difference (ITD) and interaural level difference (ILD) are the basis of the medial superior olive (MSO for ITD) and the lateral superior olive (LSO for ILD). How different inputs combine processed signals to generate parallel brainstem circuits to lead to more complex interactions is presented by the IC. Stimulus specific adaptation can increase along different inputs and indicates short-term and long-term dynamics. Obviously, SSA is an emergent functional property that converges sensory input while depressing that gets activated (Carbajal and Malmierca, 2020). Moreover, SSA is also presented in the medial geniculate body, suggesting a co-activation between the IC and MGB. Interestingly, socially relevant stimuli, such as vocalizations of conspecifics, can influence their own sources of input (Grothe, 2020). In fact, the IC may function comparably as a center for emotions and social behavior. What is clear is the tonotopic organization can be altered after an incomplete

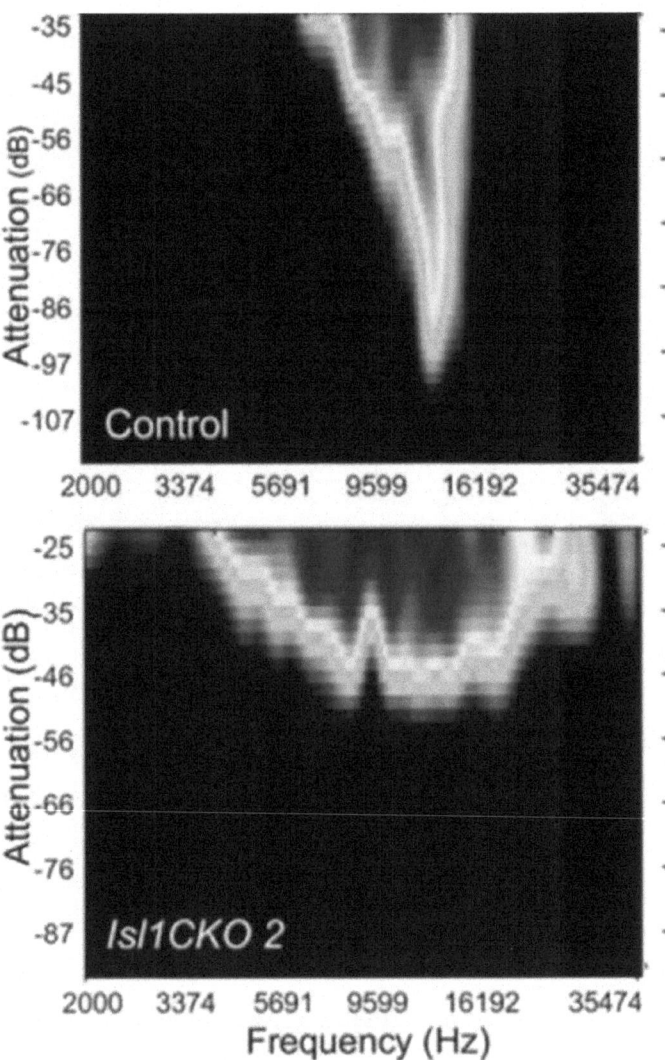

Figure 46: Two representatives tuning curves in the control (top) and the Isl1 CKO mice (bottom). Instead of showing a sharp tuning curve of 16 kHz (top) the same stimulation results in broad and irregular receptive fields. Note that flat activation compared to tuning curves. Taken from (Filova et al., 2022b).

deletion of SGNs (Filova et al., 2022b). Essentially, the tuning curves show irregular response that does not follow the typical tuning curve (Fig. 46).

5.7. The medial geniculate body (MGB)

The MGB provides, via the ipsilateral brachium of the IC, the MGB and gives ascending auditory input that reaches out the MGB to innervate the auditory cortex (Budinger, 2020). In addition, a direct input reaches the MGB from the VNC, the SOC, and the VNLL. Descending fibers come from the auditory cortex and the reticular thalamic nucleus. Three divisions are known: the ventral MGB (MGBv), the dorsal MGB (MGBd), and the medial MGB (MGBm). In addition, a suprageniculate nucleus (SG) may be a distinct division of MGBd (Malmierca, 2015). MGBv receives a tonotopic organization not found in the MGBd and the MGBm. A tonotopic connection between the MGBv and cortical auditory core, referred to as A1 and associated auditory field (AAF, rodents).

The diencephalon belongs to prosomere 2 and is defined by specific gene expression (Puelles, 2019; Puelles et al., 2012). Gene deletions show the dependence of *Shh* (Eggenschwiler et al., 2001). Prosomere 2 is just caudal to the neuropore (Fritzsch and Martin, 2022) and is separated from the zona limitans (Puelles et al., 2015) that splits between *Foxg1* (former BF-1) in prosomere 3 while the adjacent expression of *Foxd1* (former BF-2) is expressed in the ventral part of prosomere 2 (Newman et al., 2018; Puelles et al., 2015). One early gene is *Pax6* which is essential for the forebrain and also the thalamus (Manuel et al., 2022). Moreover, several human mutations of *Pax2* show major losses of auditory function. Genes that are early expressed in the developing MGB are *Wnt3*, *Tcf4*, *Meis2*, and *Irx3* (Puelles et al., 2015), among many others, some are needed for early development, such as *Zic4* and *Foxp2* (Horng et al., 2009). Moreover, *Foxp2* is considered a major gene for speech and language development (Enard, 2011). Interestingly, a late expression of *Gbx2* that early on provides the boundary of the MHB (Glover et al., 2018). *Tbr1* is expressed early on to define glutamatergic forebrain neurons. Without glutamate, the MGB has no targets and were lacking projections to the forebrain (Hevner et al., 2001). In rats, the MGB generates between E13-15 and follows a chrono-architectural development (Altman and Bayer, 1989). In mice, development starts with connections from the IC to innervate the embryonic mice between E13-18 (Fig. 47) before auditory sensory perceptions began (Gurung and Fritzsch, 2004). Targeted deletion of *Pax6*, *Foxp2*, *Wnt3*, *Tcf4*, and *Irx3* are incompletely described in their connections of MGB. Several downstream genes have been identified, including several bHLH genes (*Neurog2*, *Ascl1*, *Olig2*, and *Neurod1*) that are needed for the normal development of the MGB. Further analysis of conditional deletion

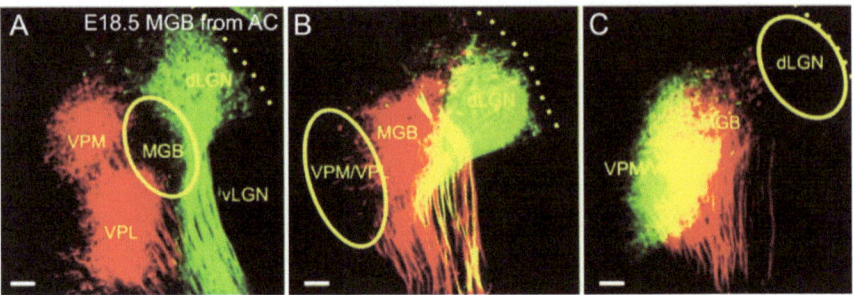

Figure 47: **Somatosensory auditory and visual projections**. Auditory cortex was labelled from dye injections in B, C, somatosensory labeling is in A, C, whereas visual projections are in A, B. Note the segregation of the different connections that reach out the MGB (B, C, red), VPM/VPL (A, C, red/green) and LGN (A, B, green). Taken from (Gurung and Fritzsch, 2004).

is needed to clarify the gene interactions in the thalamus that clarifies the various gene expression and their interactions (Dennis et al., 2019).

5.7.1. The ventral division of the MGB (MGBv)

The main input is from the ipsilateral ICc, with a small, crossed projection, which provides a tonotopic input from the lemniscal fibers to the ventro-central MGBv (Budinger, 2020). Three subdivisions have been characterized: the ventral nucleus, the ovoid nucleus, and the marginal zone. The principal neurons receive tufted dendrites to provide a dense bushy arbor. Tufted neurons receive excitatory input from glutamatergic IC and the AC. Inhibitory input is provided by GABAergic from the reticular thalamic nucleus and a few IC. Co-activation from the brachium of the IC shows an earlier input of inhibition. The MGBv is largely reciprocal connected with the AC (Fig. 45) which provides the tonotopic organization. In addition, secondary AC is reaching out to the MGBv (Malmierca, 2015). Moreover, neurons express type 2 vesicular glutamate transporter (VGLUT2 and, lesser VGLUT1) that follows a rostro-caudal direction (Read and Reyes, 2018).

5.7.2. The dorsal division of the MGB (MGBd)

The MGBd is found in the caudal and dorsal parts of the MGB for which we do not know much about its function. Several subnuclei have been identified that receive non-lemniscal input of the IC: the dorsal superficial, dorsal, deep dorsal, suprageniculate and ventro-lateral subdivisions. Its main target is the non-primary auditory cortex. Neurons are very close to the tufted neurons of MGBv and have stellate neurons to vary in density. Stellate neurons provide a larger dendritic arbor compared to tufted neurons that lack a clear orientation and are mostly non-topologically organized. Ascending fibers provide from the ipsilateral ICd and receive descending input from the reticular thalamic nucleus and secondary auditory cortical areas, the insular cortex and, the amygdala (Ito and Malmierca, 2018; Malmierca, 2015).

5.7.3. The medial division of the MGB (MGBm)

The MGBm is a flat, lentiform nucleus that contains a diverse neuronal population, some of which have long dendrites of the magnocellular neurons. Its input is from various subnuclei of the IC and also receives fibers from the SOC, the VNLL, vestibular nuclei, superior colliculus, and the spinal cord providing a multimodal integration. MGBm reaches out to all areas of the auditory cortex, somatosensory cortex, caudate-putamen, and the amygdala. Descending fibers are from the reticular thalamic nucleus and secondary auditory cortex.

5.7.4. The reticular thalamic nucleus (RTN) and the posterior paralaminar thalamic nuclei (PPTN)

The auditory sector is the RTN which is situated in the rostral and lateral surface of the dorsal thalamus. GABAergic neurons are found in the RTN and reach out to all MGB divisions. Input to the RTN comes from IC and AC. RTN may work in the global synchronization driven by wakening, arousal, sleep phases, and specific forms of epilepsy.

PPTN receives IC and projects to the thalamus from the auditory cortex and the amygdala. The prelaminar nuclei affects emotion from auditory stimuli and make a multisensory integrator from IC and SC and distribute the nuclei to the sensory cortex, associated cortex, amygdala, and striatum.

5.7.5. Integrated functional perception

MGBv deals with acoustic function, MGBd plays a role in acoustic attention and MGBm are acting in multisensory arousal and emotional auditory learning. MGBv is providing tonotopic organization to AC A1 where they can record from binaural stimulation. Tonotopic distribution starts with dorsal for low frequency while higher frequency is presented more ventrally. MGBd lacks tonotopic organization but responds to complex sounds. MGBm responds preferentially to higher frequencies despite the absence of a tonotopic map. In addition, MGBm can respond from somatosensory and auditory stimulation and the amygdala.

In summary, the MGB will receive a tonotopic organization (MGBv) and other ascending and descending inputs from another auditory cortex.

5.8. The auditory cortex (AC)

The AC receives fibers from the MGB and provides reciprocal feedback to the MGB (Fig. 45). The area of AC is surrounded by 6 (mice) to over 30 areas (humans) that deal with the associated AC (Budinger and Kanold, 2018). The topographic frequency organization is receiving a primary AC, the A1, and has several belt areas that are a secondary auditory cortex, that can present a gradient of tonotopic input (Polley et al., 2007; Polley and Takesian, 2022). Cues presented in the auditory stimulation can influence the listeners. Six distinct layers are found in about 1.1.-1.2 mm thickness to form a columnar organization (Fig. 48). Layer 1 is adjacent to the pia, layer 2 has small neurons, layer 3 has both pyramidal and non-pyramidal neurons, layer 4 are smaller neurons, layer 5 is the thickest layer that receives the larger pyramidal neurons while layer 6 contains both pyramidal and non-pyramidal neurons (Budinger, 2020; Malmierca, 2015). Most of the AC neurons are found in variable size and provided interconnections but also have spiny stellate, multipolar, chancellor, bitufted, double-bouquet, bipolar, horizontal basked neurons, and the most superficial Cajal-Retzius neurons that interact with Martinotti cells from layer 5. All pyramidal neurons are excitatory, glutamatergic projecting neurons while inhibitory interneurons are GABAergic and glycinergic.

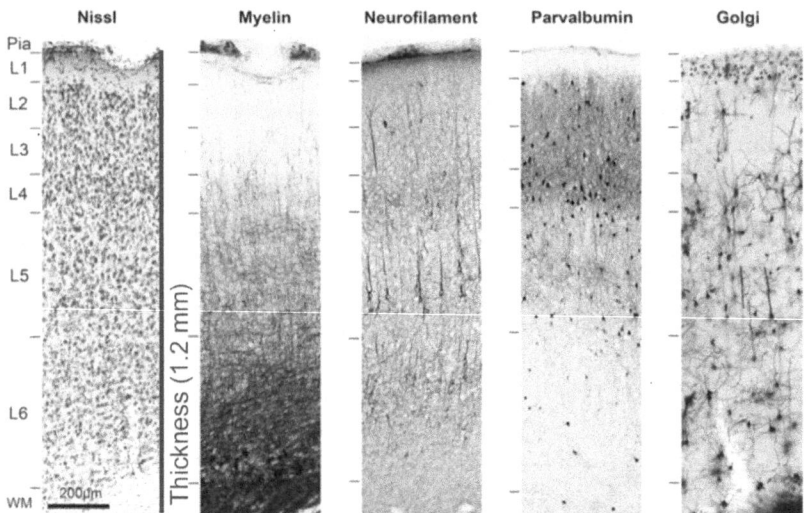

Figure 48: Six layers that characterize the AC is shown in Nissl stain (left), myelin (second image), neurofilament (middle), parvalbumin (near right), and Golgi impregnation results (far right). Note that the thickness of the cortex is about 1.2 mm. Golgi are prominently labeled in L4-6. Modified after (Budinger and Kanold, 2018).

5.8.1. Development of the AC

Evolutionary speaking, the forebrain is consistent with gene expression that is known across chordates which are expressed *Otx1, Pax6, Shh, Emx2,* and *Foxg1* (Glover et al., 2018; Kral and Pallas, 2011; Manuel et al., 2022) prior to neural tube closure of the neuropore (Fritzsch and Martin, 2022). In addition, a set of genes are needed to develop a normal forebrain such as *Dlx, Nkx2.1, Tbx1, Tbx2* and others (Arnold et al., 2008; Kempf et al., 2021; Manuel et al., 2022; Ninkovic and Götz, 2015; Puelles et al., 2015). Transcription factors vary among developing forebrain in a precise, inside-out fashion to develop reproducible spatiotemporal patterns which are essential for GRNs. The regional activation drives the production of region-specific cell types. For example, downstream of *Pax6* changes certain gene expressions such as *Ascl1, Neurog2, Neurod1, Tbr1, Sox5, Sox9,* and *Hes5* while other genes are near normal in *Pax6* deletion, including *Foxg1* (Manuel et al., 2022; Taverna et al., 2014). A possible opposing interaction of *Pax6* and *Foxg1*, both of which depend on *Shh*, and *BMPs*. A loss of *Pax6* and *Foxg1* abolishes *Ascl1* and *Olig2. Gsx2* and *Dlx1/2/3*, among others, are redirected in a different neuronal variation. A dependency is downstream of *Neurog2* that regulates *Neurod1* expression, best known is dependent on *Neurod1* for the granule neurons of the hippocampus (Dennis et al., 2019; Liu et al., 2000; Miyata et al., 1999). How many interneurons develop remains unclear after the reduction and loss of *Neurod1* null mice.

In mice, the birth of neurons is generated between E11.5-13.5 while in humans it will be much later (5-6 weeks). The earliest neurons are Cajal-Retzius neurons that develop slightly earlier [E10.5-12.5 (Dennis et al., 2019)]. Neuronal migration is adding the development of different layers (Molnár et al., 2020). Two regions are found in the preplate: the marginal zone and the subplate zone (Kanold and Luhmann, 2010). Subplate is the first input from the thalamus that will switch to direct input of cortical layer 4 in adults. The auditory cortex has 6 layers. Maturation of input from cortical neurons shows a delay in the somatosensory, visual, and auditory cortex that undergoes functional reorganizations to end up in oriented columns. The delayed innervation of cortex layer 4 after an initial innervation of the subplate neurons is a common progress in cortical excitatory ascending fibers. Most subplate neurons are lost in adults that remain as layer 6 (Goodrich and Kanold, 2020). Surviving subplate neurons may support altered circuits and such surviving neurons can cause neurological disorders (Molnár et al., 2020). Various activation is mainly glutamatergic input to the subplate and layer 4 which will receive both AMPA and NMDA input that will receive GABAergic local activity (Kanold and Luhmann, 2010). Progressive segregation from three closely related thalamus feedback (MGB, LGN, VPM) is ending up in a discrete input/output relationship between the MGB and the AC (Gurung and Fritzsch, 2004).

A major asymmetry between the left and the right side of the AC is known in humans, which typically is larger in the left hemisphere. Likewise, asymmetry is also documented for chimpanzees that have a larger left side. Bats, mice, gerbils, and rats have an asymmetry that may process species-specific vocalization. Obviously, a correlation with human language processing is certainly absent outside of humans beyond overall similarities between humans and monkeys (Budinger, 2020; Steinschneider et al., 2013).

5.8.2. AC laminar distribution and basic connections

6 layers form in the AC that have distinct inputs and outputs. The outermost Layer 1 (L1) primarily consists of corticocortical axons and connects with apical dendrites of pyramidal neurons and various interneurons of unique interneurons: Horizontal, GABAergic make up 90% of inhibitory neurons and Cajal-Retzius are about 10% of excitatory neurons that are parvalbumin positive to interact with dendrites of pyramidal neurons. L2 consists of small and medium pyramidal neurons and a large range of interneurons that spread between L2-L5/6 (see below). No information correlates with the anatomical features of pyramidal neurons. L3 pyramidal neurons have complex dendritic trees that connect intrinsic, bilateral cortical, and thalamic connections. 24% are GABAergic of interneurons. L4 is almost devoid of pyramidal neurons (compared to the visual and somatosensory cortex). Spiny stellate, excitatory neurons are dense in some mammals and can be fewer neurons in others. L5 are particularly large pyramidal neurons that extend across up to L1 and reach out to subcortical and local cortical areas, including the corticofugal projections that reach out to the thalamus, midbrain, and brainstem. The larger pyramidal neurons are unaccommodating and nonadaptive, bursting or regular spiking neurons, some of which are inverted pyramidal neurons. The MGBd bear giant boutons that likely provide feedforward activation. About 27% of interneurons are found in L5. The most diverse is L6, consisting of small to large pyramidal neurons that reach out the thalamus (MGBv). Mostly GABAergic interneurons make up 16%, the lowest ratio of pyramidal neurons. Persistent subplate neurons may consist that could activate under unusual conditions (Kanold and Luhmann, 2010; Molnár et al., 2020).

Thalamocortical excitation inputs are broad while intracortical excitatory inputs have much sharper tuning properties. Inhibitory inputs are balanced to excitatory inputs that have similar best frequencies. The interneurons consist of several mostly GABAergic, inhibitory inputs among the neuroglia form, chandelier, basket, bipolar, bitufted, double-bouquet, Martinotti and glutamate spiny stellate, and Cajal-Retzius neurons (Budinger and Kanold, 2018). The distribution is variable across L2-6 with different concentrations (Budinger and Kanold, 2018).

Obviously, the AC circuits have a bewildering complexity that shows a tonotopic organization of frequency preferences. The AC can extract meaningful signals for conspecific vocalizations. In rodents, frequency presentation is tuned to respond about 350 µm away to a particular tonotopic location that may follow the minicolumnar organization of the AC of about 40-50 µm (Budinger, 2020). Cajal-Retzius neurons are in L1 and act as transcolumnar, which will also form a transcolumnar organization in spiny stellate cells in L3/4 while interneurons are columnar or a few columns wide that interact via gap junctions.

5.8.3. Corticocortical and corticothalamic connections

Thalamic afferents come from different subdivisions of MGB that are in contact with pyramidal and interneurons. For example, the MGBv contact with tonotopic input ends up in larger boutons reaching A1 and AAF in L4 and L3. In addition, an overlapping non-tonotopic pathway will reach from MGBd in L3/L4 whereas MGBm largest boutons are in L6 and L1 (Fig. 49). Notably, all layers receive some different sizes of boutons between MGBv, MGBd, and MGBm.

Outputs of pyramidal neurons can connect to different layers within and across the cortex. The short intracortical afferents end up in L2 and L3 which provide feedforward inputs. Feedback inputs from pyramidal neurons reach out to L1, L2, L5, and L6 showing a distinct set of interactions while lateral pyramidal neurons reach out to all layers (Fig. 49). Interhemispheric connections across the corpus callosum come from L3 and L5 to innervate the contralateral AC. The output of corticothalamic fibers reaches MGB and reticular thalamic input from L6 whereas L5 neurons extend to innervate the MGBm, MGBd and IC. In addition, from L5 and L6 come corticofugal output that reaches out to the IC and other areas of the brainstem.

The thalamic cortex will have in every mammal an A1 that is tonotopically organized. In addition, up to 30 distinct layers of auditory input are characterized by the AC. For example, in the macaque monkey about 13 layers receive next to A1 and several core inputs have also been a belt and a parabelt distinct inputs and various interactions that are connected in a hierarchical fashion that may cross-connections among each other (Budinger, 2020): core fields have strongest connections with each other and surrounding belt fields, belt fields have strong connections with an adjacent belt but also has connections with the core and the parabelt fields while parabelt fields interact with each other and adjacent belt fields. In humans, a core, belt, parabelt, and higher order areas can be demonstrated by MRI techniques. Overall, hierarchical, and parallel processing organization is found in the auditory cortex.

A massive corticothalamic provides descending connections with the MGB, IC, and brainstem, including the olivocochlear system. Descending

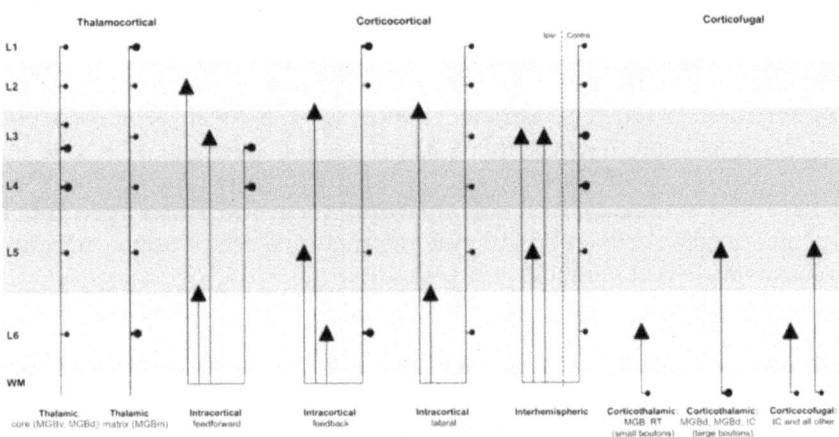

Figure 49: The input from thalamocortical input (left), and the output from corticocortical (middle) and corticofugal (right) is shown. More prominent inputs are indicated as larger dots while weaker inputs are shown as small dots. Note the corticocortical connections are in part ipsilateral while some have a contralateral connection from L3 and L5. Corticofugal output is provided by L5/6 large and medium sized pyramidal neurons. Taken from (Budinger, 2020).

pathways act dynamically to response sensitivity and selectivity of auditory neurons that depend on the behavioral and sensory context from which sounds are heard (King, 2020). Layer 6 projects predominantly to the MGBv and the reticular thalamus (RT) while Layer 5 reaches out to MGBd, the shell of the IC, and the amygdala. Physiological connections are provided to be altered by sound from AC activation that affects the lateral lemniscus (LL), the superior olivary complex (SOC) and the cochlear nuclei [CN (Liu et al., 2019)]. Moreover, corticofugal pathways affect context-dependent auditory processing (King, 2020). The amygdala is in the social interaction that integrates auditory stimulation with vision, somatosensory, gustatory, and olfactory sensory input (Wenstrup et al., 2020). Stimulation reaches out to the amygdala from the associated AC and the MGB, mostly from MGBm but also from the MGBd and the SG. The AC receives amygdala input and will also reach out to the MGB.

Primary sensory cortices are typically considered a selective input to A1 (AC), S1 (somatosensory), and V1 (visual) connections but it is now shown that multisensory responses of the A1 are documented in several mammals. Auditory information deals with the identification of the sound source ('what') that processes along the rostro-ventral cortex while other information is concerned with the spatial location of the sound source ('where') which is distributed along the caudo-dorsal cortex (Rauschecker, 2020). It also depends on the emotional status that modifies the response to the communication sounds (Wenstrup et al., 2020). What needs to be documented is the tonotopic, non-tonotopic, and multimodal hierarchy systems that interact with them.

5.8.4. Cortical plasticity: From the ear to the AC

Cortical plasticity goes beyond the changes during development (Patton et al., 2019). Here, I will present four different losses of cochlear deletion or loss and how it correlates with the effects of the auditory system:

(A) The deletion of *Neurod1* (and *Isl1*) causes reorganization of the tonotopic organization.
(B) The deletion of otoferlin leads to impaired cochlear synaptic transmission.
(C) Deletion of type I SGNs with ouabain can demonstrate the remaining auditory processing.
(D) Deaf subjects can reconnect fibers to reach AC input.

(A) The role of *Neurod1* was initially described and showed the near complete loss of SGNs that have no proper activation of the AC, revealed with auditory evoked potentials (Kim et al., 2001). A follow up showed the central projection of SGNs that contact the cochlear nuclei that have an overlap of vestibular and cochlear neurons (Jahan et al., 2010). What is unclear is how many granule cells develop in the cochlear nuclei, the IC, MGB, and AC that are highly expressed for *Neurod1* (Gurung and Fritzsch, 2004; Liu et al., 2000; Miyata et al., 1999). It is possible that the absence of *Neurod1* can result in the absence of cochlear evoked potential revealed in the previous work. Essentially, we require the connection from the MGB to the AC and document the absence of specific inhibitory neurons in the IC, MGB, and AC (Budinger, 2020). A conditional deletion of *Neurod1* using *Isl1-cre* reduces the remaining SGNs to have an overlap of central projections (Macova et al., 2019). Moreover, a 40% reduction in cochlear hair cells was accompanied by a reduction of SGN to about 20%. Since *Isl1* is not expressed in the cochlear nucleus, the reduction of volume to about 60% is likely due to the reduction of SGN central projections. There was no volume reduction in the IC that is highly expressed *Neurod1* (Gurung and Fritzsch, 2004). ABR and DPOAE are affected by conditional deletions and show abnormal tuning curves. Most importantly, the isofrequency presentation shows a broad central cochlear and IC input that lacks a single best frequency in the tuning curve (Macova et al., 2019). Using *Neurod1-cre* to eliminate *Isl1* (Fig. 46) is showing a nearly complete absence of tonotopic organization in the IC (Filova et al., 2022b). Fewer SGNs develop and migrate into a different position in the modiolus reaching into the cochlear nucleus with incomplete overlapping projections, lacking a tonotopic central organization of the CN and IC. Either deletion of Isl1 or Neurod1 requires a proper investigation for MGB and AC auditory input.

(B) Otoferlin (*Otof*) is needed for ribbon synapses to facilitate vesicle fusion and exocytosis (Michalski and Petit, 2019; Moser, 2020). Otoferlin null mice are reducing vesicular release and are deaf after P21 old mice (Roux et al., 2006) and can provide progressive hearing loss in humans (Mukherjee et al., 2021). At P11, WT and KO performed similarly while P38 old KO mice showed impaired synaptic maturation that reduced MNTB excitability (Müller et al., 2022). In the absence of otoferlin, they show a reduced activity that results in imprecise topography of the MNTB and LSO connections (Müller et al., 2019) which may impact ILD coding. Recording from earliest growing MGB derived projections reaches out first to the subplate neurons that develop exuberant connections, both excitatory and inhibitory neurons (Mukherjee et al., 2021). Overall, otoferlin KO mice result in impaired hearing loss that has sculpted cortical connections.

(C) Losing SGNs with age is declining availability to drive fewer fibers, like with cochlear implants (Elliott et al., 2022a; Wu et al., 2019). An extreme approach takes 95% of type I neurons with ouabain, remaining only type II SGNs that innervate the OHCs (Barkat et al., 2011; Chambers et al., 2016; Lang, 2016). Using a sparing of type I SGNs bilaterally shows a recover ABR of 70 dB sound pressure level (SPL) was reduced by 70% and showed an elevated ABR of 30 dB (Polley and Takesian, 2022). Sound-evoked responses can activate in the AC but differentially drive the inhibitory and activates the excitatory pyramidal neurons that are subsequently amplified by disinhibition which results in hyperactivity and increased neural gain while it also reduced the background noise. In summary, internal cortical noise can underly perceptual difficulties after incomplete deletion of type I SGNs.

(D) In cats and humans, deaf subjects can show a cross-modal reorganization that can lead to novel connections from visual and somatic sensory input into the AC (Kral and Pallas, 2011). Forced rerouting, after the IC and SC removal, can respond to visual stimulation that may activate the AC by providing complex spatial visual input via the retinal-MGB-AC (Kral and Pallas, 2011). In the cat, hearing activity is simultaneously in all layers while the cat deaf animals show reduced activity in the deep layers which suggests it's compromised. The absence of hearing interferes with the functional microcircuit in the AC which results in reduced supragranular layers (Kral et al., 2017). Corticocortical interactions may develop in a late loss in humans, including cochlear implants (Kral et al., 2019). Unfortunately, little information is presented about second order connections that could relate to cross-modal activity that could swap modality in congenital deafness cats and humans. It should be noted that the auditory system has input into a vast area of the non-auditory cortex (Fritzsch et al., 2022; Rauschecker, 2020).

5.8.5. Auditory cortical processing revealed

Previously, Merzenich, Kaas, Mishkin, and Rauschecker's work documented the auditory cortex, particularly in monkeys and humans. Primary sensory cortices of A1 are distinguished by receiving their main input from the MGBv which shows the best frequency along the lateral sulcus in the rhesus monkey (Rauschecker, 2020). Picking out particular sounds is key for the auditory system to better understand how to recognize speech in noise, a major challenge for older people, in particular (King and Walker, 2020). The auditory system is driven by sound sensing but is also affected by the interpretation of the acoustic environment. Various cortical neuron types can be related to the position of the A1 (Malone et al., 2020). Only about 18% of A1 from MGB also receive other cortex inputs as well as basal ganglia, amygdala, and hippocampus. Its function is to receive convergence of subcortical and corticocortical information to act between the auditory system and sensing, perception, and interpretation of acoustic information for processing (Kanold et al., 2014). Primary auditory cortex is necessary for sound awareness and perception while their role will be required for detection, discrimination, categorization, and task-specific decision making (Malone et al., 2020). Overall, the primary AC is one node of an extended loop of processing rather than a unidirectional information flow (Polley and Schiller, 2022; Winer and Schreiner, 2011).

The receptive field of a given neuron requires a particular stimulus dimension such as spectral pitch or frequency, intensity or loudness, periodicity, spectral envelope or timbre, and sound localization or interaural time and level differences (Malone et al., 2020). Stimulus properties are often varied simultaneously and independently to assess their interactions on a response. The frequency response area (FRA) provides tone frequency and intensity. What makes it so difficult to characterize A1 receptive fields is that it makes it problematic to predict the response to natural sounds (Carbajal and Malmierca, 2020). Furthermore, neuron types may be distinguished based on electrophysiological and neuroanatomical features or immunochemistry and anatomical reconstruction after intracellular labeling or the use of molecular or genetic techniques such as fluorophores. In many cases, one can generate a close but not perfect agreement with different methods.

Most A1 neurons show a single peak of FRA that characterizes a single region of intensity centered on the constant frequency (CF). However, some neurons may have multipeaked tuning curves. Responses to sound will depend on the recent history of stimulation can suppress sound processing in nearby excitatory neurons while inactivating other neurons may alter the response bandwidth (Malone et al., 2020). Moreover, spectral bandwidth specificity may aid in the extraction and evaluation of complex spectral shapes. Searching for the best responses within a multi-dimensional parameter may need more advanced methods to study neurons to encode a range of different stimulus aspects. Receptive field analysis provide insights between subcortical and cortical processing aspects while limiting the ability to fully identify the neuronal properties in response to natural sounds. For example, range adaptation can be associated with adjustments in human hearing while sound level discrimination thresholds vary with stimulus contrast to affect the contrast gain control. We know that electrocorticography (ECoG) data from neurosurgical patients listening to speech in the presence of abruptly changing background noise have shown that auditory cortical neurons rapidly adapt to the noise, resulting in enhanced neural coding and perception of the phonetic features of speech (King and Walker, 2020).

Specific contributions of mammalian A1 require different sensory and perceptual tasks that will need to become a more holistic approach to be fully understood. Thus, sound offsets, generated through the dorsal cochlea, superior olivary nucleus, will remain temporally precise interaction of the MGB. Neurons of the MGB form offset-encoding synapses to reach the auditory cortex (A1) from those sound onsets. In addition, the auditory cortex makes inhibitory inputs to sharpen the temporal responses which highlight the beginning of an auditory event that processes a complex auditory scene. Complete knowledge is a solution to help with the cocktail party problem that will it have the ability to help hearing and communication, particular in the elderly. Unfortunately, the precise purpose of A1 within the auditory pathway remains somewhat obscure (Malone et al., 2020).

CHAPTER 6

The Evolution of the Auditory System as a Blueprint for Sensory Expansions Using Established Principles – Evolving New or Adding Existing Senses to Expand the Perception of the World

The brain is the central part of the auditory system that extends from the cochlear nucleus to the telencephalon in gnathostomes. Lampreys have a similar connection from the visual and trigeminal connections (Suryanarayana et al., 2020) to the telencephalon that likely will not have an auditory system that could have some sound perception using the common macula. How can we understand the auditory system that derives in part from the previous lateral line and electroreception after that input was absent in amniotes and frogs? I will provide an overview of common and derived inputs first followed by the telencephalon that receives the auditory input and beyond.

6.1. Evolving the auditory central projection revealed

We already know that virtually all vertebrates, with or without a lung to help sound perception, may hear at a lower sound detection and a limited frequency, probably including lampreys, elasmobranchs, lungfish, and certain lungless salamanders (Capshaw et al., 2020). What sets apart is the impact of the basilar papilla that is common in all sarcopterygians, starting with *Latimeria* that builds up a redundant endorgan (Fig. 50), unrelated to vestibular sensory input, which eventually becomes the basilar papilla that will evolve into the organ of Corti in mammals. Actinopterygians had a parallel transformation that associates in different ways with the lungs to allow much higher frequency that can be perceived (Schulz-Mirbach et al., 2019). Sound perception without a lung connection may be limited to about 250 frequency range hich requires a very high intensity to drive hearing input (Capshaw et al., 2022). Higher frequency

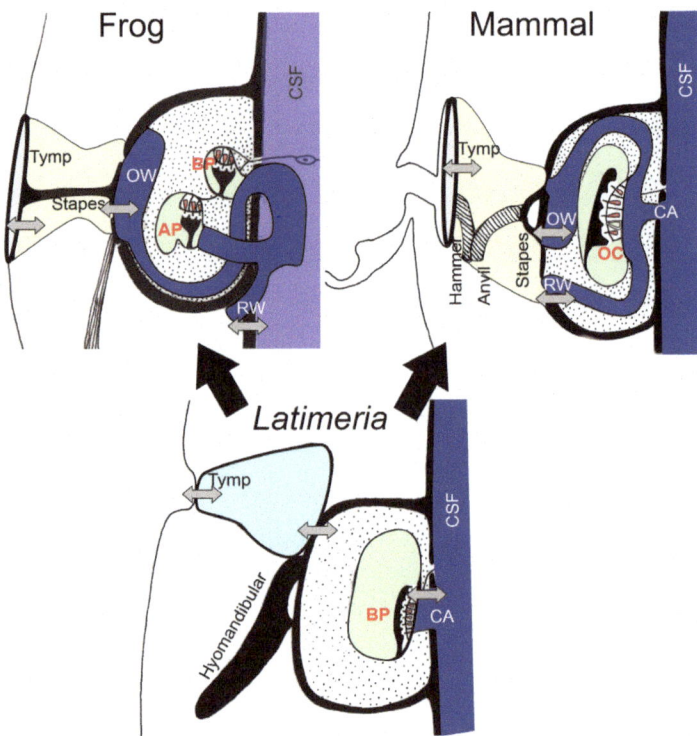

Figure 50: Formation of *Latimeria* compared to frogs and mammals. *Latimeria* has a cochlear aqueduct (CA) adjacent to the basilar papilla (BP) comparable to mammals. Note that the CA is replaced with a perilymphatic space next to the basilar papilla (BP) in amphibians that has a convoluted connection between the oval window (OW), the amphibian papilla, and the round window (RW). The scala vestibuli next to the oval window in mammals passes the scala tympani to end up in the round window (RW). The stapes (frog) and three ossicles (mammal) connect to the tympanum (Tymp). A tympanum exists in *Latimeria* but likely does not have air but is driven by the fluid across the cochlear aqueduct, the ear, and cartilage from the inside-out (from the basilar papilla/cochlear aqueduct to the tympanum). In contrast, frogs and mammals have a sound driven by air next to the tympanum which will eventually be converted to endolymph to provide sound perception from the outside-in (from the tympanum to the oval/round window). Modified after (Fritzsch et al., 2023b).

and lower intensity happened after the evolution started in *Latimeria* which has an identical connection with the basilar papilla that drives the cochlear aqueduct (Fig. 50) from the inside-out, from the perilymph via the basilar membrane. What happened next is the evolution of the tympanic membrane that evolved independently four times: Latimeria, most frogs, sauropsids, and mammals (Fig. 50). Most tetrapods have a transformation of the hyomandibular bone into the stapes that adds into the opercularis in amphibians (Jaslow et al., 1988). How the systems work for sound in *Latimeria* is unclear but shows a disruption of the basilar papilla after pulling them up from water that causes decompression, comparable to human barotrauma. We can assume that the basilar membrane was working much like tetrapods that evolved early but was lost in several species (lungfish, most gymnophionans, and many salamanders). Hair cells may be larger in *Latimeria* but become a very small number of hair cells in amphibians and certain sauropsids while others have over 10,000 hair cells in the owl and the whales (Carr, 2020; Manley, 2018). The length of the basilar papilla is short in amphibians, elongate in sauropsids, and can become 100 mm long in mammals. Best described, sauropsids and mammals differ in the organization of hair cells: while hair cells are consisting of tall and short hair cells in most archosaurs, it forms separate inner hair cells and outer hair cells in mammals (Manley, 2018). Moreover, monotremes have multiple rows of inner and outer hair cells whereas therapods have a single row of inner hair cells that has about 95% of all spiral ganglion neurons while three rows of outer hair cells are innervated by just about 5% of SGNs (Fig. 40). The pattern of the basilar papilla is different in amphibians that have a broad innervation without any clear difference in hair cells (Fritzsch and Neary, 1998; Smotherman and Narins, 2004). A tectorial membrane is common among tetrapods that have the basilar papilla in sarcopterygians but have the additional membrane from the amphibian papilla (Simmons, 2019). How the system worked in *Latimeria* may have started at the cochlear aqueduct uses for low frequency driven by the intracranial joint, comparable to the size of the cochlear aqueduct in certain birds that may have tuned specifically for low frequency (Zeyl et al., 2020). Essentially, the basilar membrane works from inside-out (*Latimeria*) to generate pressure using the intracranial points (Fig. 50) whereas in most amphibians, sauropsids, and mammals they worked from the outside-in, starting on the tympanum to release the sound pressure via the round window (Fig. 50).

The basilar membrane endorgan of the organ of Corti is responding from the apex driven by the lowest frequency of the tonotopy of sound, whereas the highest frequency follows at the base, with the exception of inverse tonotopic organization in certain lizards (Manley, 2018). In contrast, amphibians can distinguish between two auditory endorgans, the basilar papilla (50 or fewer hair cells) and the amphibian papilla (50-150 hair cells), the latter may evolve from the neglecta papilla (Fritzsch and Wake, 1988). The frequency range is limited to under 8 kHz in most sauropsids while mammals have a much higher frequency range of 20-40 kHZ that can expand to over 100 kHz in bats and whales. In contrast, the basilar papilla in frogs responds to 2 or more kHz while the amphibian papilla responds to sounds from 300-1.800 Hz. Amphibians show a different organization from the basilar and amphibian papilla on a short recess while it forms a single set of basilar papilla/organ of Corti in amniotes. Unfortunately, we do not know the frequency range that can respond to the basilar papilla in *Latimeria*. What needs to be explored is sensory organ connections between the basilar papilla with specific neurons to reach out to the cochlear nuclei (Grothe et al., 2004).

Central projections are identified in a tonotopic innervation in frogs, birds, and mammals that ends up in the auditory (cochlear) nucleus. Frogs have a tonotopic, central projection from the amphibian papillae that end up in the dorsolateral nucleus (Carr, 2020; Fritzsch and Neary, 1998). No tonotopic projections is found in the basilar papilla in frogs. There should have some incompletely segregated from the basilar and amphibian papilla in salamanders (Manteuffel and Naujoks-Manteuffel, 1990). Basilar papilla fibers in *Latimeria* have not been traced and are likely absent in lungfish, most gymnophionans, and salamanders. A binaural commissural interaction between the two sides of the hemisphere could play a role in directional hearing in frogs that show about six types of cochlear nuclei (Feng and Schellart, 1999). The number of different kinds of connections will be increasing to end up with 12 or more cochlear nuclei neurons in mammals. Sauropsids have a distinct population, the angular nucleus (r2-4), that has a different input to receive the magnocellular nucleus segregated from one nucleus (r5-r8). Mammals have a ventral cochlear nucleus (r2-4) and the dorsal cochlear nucleus (r4-5). A commissural connection exists in mammals (T-stellate neurons) that seems to be also in the chicken (Lipovsek and Wingate, 2018). In addition, granule cells derived from the eminentia granularis in gnathostomes transform into granular neurons after the loss of lateral line and electroreception in amniotes.

Higher connections in frogs have a contralateral input from the cochlear nucleus to innervate the superior olive, the lateral lemniscus, and the inferior colliculus (equivalent to the torus semicircularis) from which they have tonotopic input that can have at least three subnuclei (Feng and Lin, 1991). The migration of the superior olive and the lateral lemniscus has not been documented in its origin. A similar distribution of superior olive is found in chickens that have an additional dorsal origin of the laminaria nucleus (Lipovsek and Wingate, 2018). The origin of the lateral lemniscus requires further studies. The inferior colliculus in sauropsids reaches out the equivalent of the inferior colliculus where it receives a core and a shell that shows a tonotopic input in certain areas in frogs and birds (Carr, 2020). In mammals, several distinct populations of the superior olive are known that are migrating from r5 and provide a minor contribution from r4. Not all neurons migrate from the *Atoh1* dorsal region which requires additional study to define the origin. The lateral lemniscus is in part migrating from r4 and in part, it is unclear where they migrate from in mammals. A core and shell are characterized in the mammalian inferior colliculus and a tonotopic input converges to become the ascending cochlear projection (Fig. 45).

In contrast to the new formation of the cochlear nucleus in tetrapods, teleosts have a different sound input that ends up in the same inferior colliculus [named torus semicircularis (Walton et al., 2017)]. No formation of the superior olive is formed in teleosts (Bleckmann et al., 1991), some of which may have both electroreception and lateral line input that shows the parallel central projection in gnathostomes (Carr, 2020; Engelmann and Fritzsch, 2022) and, possibly, lampreys (Fritzsch, 1998).

Ascending projections to the thalamus can be found in gnathostomes that have an equivalent MGB in teleosts, amphibians, sauropsids, and mammals (Grothe et al., 2004). How similar these neurons have additional hypothalamic connections in certain gnathostomes remains unclear. However, the main ascending input to an MGB nucleus is common in frogs, sauropsids, and mammals. No tonotopic organization is found in the forebrain of frogs (Feng and Schellart, 1999). Despite the different organization of the cortex in teleosts, frogs and birds have an equivalent to the auditory cortex in mammals that are known to end up as tonotopic input in birds and mammals (Carr, 2020). The projection from a core MGB to reach out the cortical A1 is common in mammals but has additional secondary connections, the belt, that can increase the number of the distinct auditory organization from fewer areas that can increase in 30 or more subdivisions in the humans (Budinger, 2020; Malone et al., 2020).

In summary, a simple superior olive and lateral lemniscus in frogs become a more complex origin in chicken and in mammals that adds a unique migration from the lateral lemniscus from r4 in mammals. Tonotopic input is known in frogs' amphibian papilla, birds, and mammals that reach out from the cochlear nucleus to the equivalent of the auditory cortex. It is adding in mammals a unique contralateral input from the MNTB that combines with ipsilateral LNTB that provides, with MSO, for interacting with the tonotopic frequency as well as interaural time difference (ITD) and interaural level difference (ILD). ITD plays a role in sound processing in sauropsids but has limited interconnections in frogs. ILD will be used in small mammals to detect the direction of sound in humans with a frequency of >2 kHz. Larger mammals, including humans, use for low frequencies below 2 kHz to determine ITD directionality which is the difference in the arrival sound waves of two ears (Pecka and

Encke, 2020). In addition, we need to have yet another way to discriminate among sound directions using possibly interaural phase differences (IPD) that can be used from the phase delay between two monaural responses. The inferior colliculus receives all information from the cochlear nucleus, superior olive, and lateral lemniscus which likely will provide a wide range of representation. What remains to be seen is somewhat unclear information on the spatial representation of generated local range (Kopp-Scheinpflug and Forsythe, 2018). Further, we need to add different temporal information at the timescale range to extract sound features such as amplitude modulation in the inferior colliculi, the MGB, and the auditory cortex (Kopp-Scheinpflug and Linden, 2020).

6.2. Auditory sound processing

Human sounds and emotions will be conveyed by speech, pitch, loudness, timbre, speech rate, phrasing, and non-verbal vocalizations. Only humans have this information that can have a unique processing of sound (Rauschecker, 2020; Rauschecker and Scott, 2009). Evolution of spoken language must have emerged from neural mechanisms in animals and the capacity for vocalization is ancestral to tetrapods (Fitch, 2018). We can add various vocalizations, including a large range of mammals, birds, certain lizards, and frogs. Moreover, many fishes, frogs, birds, and mammals generated sound in various ways (Bass et al., 2015; Fitch, 2018).

In addition, humans sing alone and in groups. While singing can be human specific but others also do, whales 'sing' by producing variations of sounds in that (Schneider and Mercado III, 2019). Moreover, the whales have the same range in size compared to humans (Herculano-Houzel, 2016). In addition, singing exists with specific birds, like the nightingale, which can change its song frequently. The human larynx and vocal tract are shared with other mammals. Incidentally, in humanoids, lose a vocal sac that is present in chimpanzees (Fitch, 2018). The avian syrinx is a unique source organ representing a novel way to sing in birds that was recently demonstrated to have the same sound production across mammals and birds (Elemans et al., 2015). Likewise, frogs produce sound from the larynx (Ryan and Guerra, 2014). In contrast, bony fish have several ways to produce sounds (Bass et al., 2015). Among many sound producers, the most interesting is the siluriforms that have both sound and electric discharge showing a different input (Bleckmann et al., 1991; Kéver et al., 2021; Schulz-Mirbach et al., 2019). Land vertebrates use sound in air and the sound production somewhat differs depending on the hydrodynamic function (Engelmann et al., 2000). Interestingly, the sound production may be from the same motor neuronal input to sound in fish, frogs, birds, and mammals (Bass et al., 2015).

Upstream of brainstem motor neurons are direct input to activate sound production by higher orders such as the cortex in mammals and certain sauropsids, including birds. It is unique that sound vocalization in humans can be activated unlearned, innate vocalization (coughing, for example) shared with many amniotes, but which have the ability to learn a novel language in humans (Fitch, 2018; Savage et al., 2020) that has a similarity between human and monkey (Steinschneider et al., 2013).

Before learning anything, we need to understand the origin of words: the infants' early canonical babbling. We know that infants can produce language-like consonant-vowel syllables that affects speech. The types of consonants and syllable structures most frequently produce in vocalizations at the babbling stage and are those used most often in children's earliest words such as baba, mama, papa, nana, etc. (Keren-Portnoy et al., 2009). Consolidating the infant babble requires a concise approach that helps to understand how language develops (Daffern et al., 2020). Once the first understanding of words is made by an infant, the next steps can be built on the babble level. Infants of age 2-3 can speak and sing (Savage et al., 2020). The biggest question is: "Why study the origins of music, language, or any other human behavior? It's unlikely that anyone will ever explain the full extent to which a particular behavior is accounted for by one or more adaptations because, given its complexity, human behavior cannot be exhaustively measured (Mehr et al., 2020)". How the evolution of musicality could generate gene culture coevolution. We know that proto-musical behaviors could initially arise and spread as cultural inventions that had feedback effects on biological evolution, likely because of their impact on social bonding. Two major articles (Mehr et al., 2020; Savage et al., 2020) come to contradictory explanations for music's evolution. Both articles minimize the sense in which musical traits may be explained as byproducts of non-musical adaptive functions. In addition, both articles present unified theories purporting to explain the evolution of musicality in terms of music's adaptation. The continued vocabulary explosion may depend on association learning, familiar word recognition, and logically distinct associative learning (Kucker et al., 2015; McMurray et al., 2012).

Obviously, infants respond to vision early on, but it takes at least 1-2 years before language starts. Why does it take so long to speak when hearing already starts before birth? How do these features change with infant age and differ across languages? We understand that vocalizations may induce positive emotions through raised pitch and pitch variability and modulated loudness reflecting speakers' positive valence and heightened arousal (Cox et al., 2022). Interestingly, even infant-directed songs in a foreign language can induce relaxation in babies, suggesting a common song effect across a range of the world. How learning in infants younger than one year remains somewhat unclear despite the progression of infant-directed songs and words. Simply speaking, we have about 1-2 years of a newly born baby to begin speaking and singing.

In the adult monkey and man, two streams exit from the auditory cortex: the caudal belt to reach the dorsolateral prefrontal cortex while the surrounding belt projects to the ventrolateral prefrontal cortex. Overall, the auditory ventral pathway role plays in perception and is largely consistent with a 'what' pathway, whereas the dorsal pathway takes on a sensorimotor role involved in the action ('how'), including spatial analysis [Fig. 51 (Rauschecker and Scott, 2009)]. Thus, speech processing of an antero-lateral gradient in which the complexity of preferred stimuli increases, from tones and noise bursts to words and sentences, while a posterior-dorsal stream provides the spatial role and motion detection.

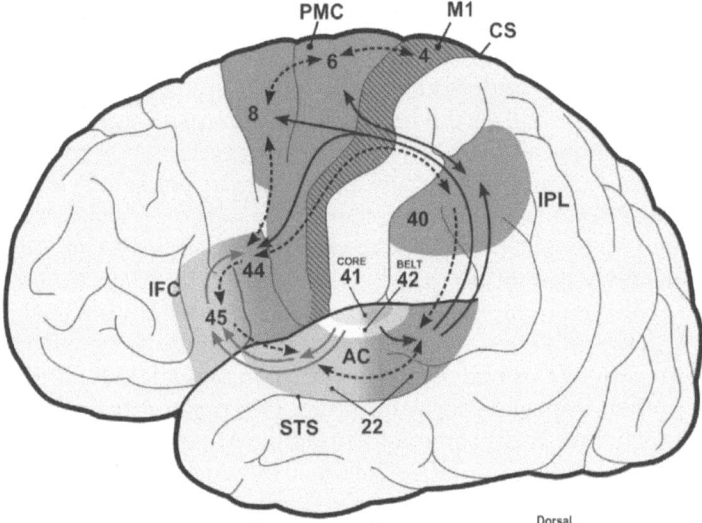

Figure 51: Dual auditory processing scheme of the human brain and the role of internal models in sensory systems. This scheme closes the loop between speech perception and production and proposes a common computational structure for space processing and speech control in the postero-dorsal auditory stream. Antero-ventral and postero-dorsal streams originates from the auditory belt. The postero-dorsal stream interfaces with premotor areas and pivots around inferior parietal cortex, where sensory event information with a predictive efference of motor plans. A forward mapping is object information, such as speech, which decoded in the antero-ventral stream including inferior frontal cortex (area 45) and motor-articulatory representations (area 44, ventral PMC), whose activation is transmitted to the IPL as an efference copy. An inverse mapping will attention- or intention-related changes in the IPL that influence the selection of context dependent action programs in PFC and PMC. AC, auditory cortex; STS, superior temporal sulcus; IFC, inferior frontal cortex, PMC, premotor cortex; IPL, inferior parietal lobule; CS, central sulcus. Numbers correspond to Brodmann areas. Taken from (Rauschecker and Scott, 2009).

Multiple parallel input modules are advocated for the dual stream model (Fig. 51).

In a detailed analysis, auditory cortical processing was shown to play a role in 360 cortical areas of 171 human connectomes at 7T using diffusion tractography for imaging showing the functional connectivity between brain regions. A hierarchy was documented of auditory cortical processing that was expanded from A1 to belt regions. Consistent with a 'what' ventral auditory stream which are language-related Broca's area (area 45). Likewise, a 'where' stream from the superior parietal region may form a language-related dorsal stream to reach out area 44 (Rolls et al., 2022). A1 has strong input from the MGB but also has connectivity from the visual (unidirectional) and somatosensory regions. A weak connectivity was shown for the cingulate area. Area 52 is unique to humans and receives a strong input from the MGB, which is weaker compared to A1 which also connects somatosensory and insular regions. The belt input is reciprocal to A1. A hierarchy is A1 which is stronger compared to belts and 52 and the weaker is A4 followed by A5 (Rolls et al., 2022). Information flow from areas 45 and 44 can be part of a route from semantic to language output in speech production and articulation (Rauschecker and Scott, 2009).

In summary, the cortical human connections are between the A1 to reach out, through multiple steps of belts and area 52, to area 45 and 44. How the earliest infant can have connected these specific connections to allow the output to vocalizations remains to be seen.

6.3. The auditory cortex reaches out beyond the A1

The auditory cortex (A1) is a central hub located at a pivotal position within the auditory system, receiving tonotopic input from the MGB and playing a role in the sensation, perception, and interpretation of the acoustic environment (Malone et al., 2020). 'Receptive fields' (RF) end up with a particular stimulus dimension: the choice of acoustic stimulus dimensions to explore is related to the basic perceptual attributes such as spectral pitch (frequency), loudness (intensity), periodicity, or virtual pitch (amplitude modulations, harmonic series), timbre (spectral envelope), and sound location (interaural time, level differences, spectral shape). In humans, hemispheric asymmetries between the left and right auditory cortex form in gross anatomical features and cortical microcircuitry (Budinger, 2020; Polley and Takesian, 2022; Rauschecker, 2020).

There are two major classes of neurons in the neocortex: principal cells and interneurons. The auditory cortex is organized into six horizontal layers, another defining feature of the neocortex. It has anatomical and functional vertical columns and intense interhemispheric connections between the auditory cortices of both hemispheres, which cross the midline via the (caudal body of the) corpus callosum (Budinger, 2020; Malone et al., 2020). Primary sensory cortices like A1, S1 (somatosensory), and V1 (visual) are not purely unimodal but also process other non-matched sensory and non-sensory information. Projections from V1 and, to a lesser degree, from S1 towards A1 arise from subgranular layers and provide feedforward organizations (Budinger, 2020).

The auditory corticobulbar system includes projections from the nuclei of the medial geniculate body (MGB), the inferior colliculi (IC), the lateral lemniscus (LL), superior olivary complex (SOC), and the cochlear nuclear complex (CN) including the medial olivary afferents that project out to the OHCs (King, 2020; Simmons et al., 2011). The prominent corticothalamic, corticocollicular, colliculofugal, and olivocochlear connections descend from auditory pathways that include long-range connections (Budinger, 2020; King, 2020; Malmierca, 2015). Overall, the feedback loop might modulate auditory response properties of neurons in the midbrain and hindbrain, leading to an adaptation of their sensitivity to sound frequency, intensity, and location (Budinger, 2020; King, 2020). The AC interacts with multiple other sensory projections.

Auditory-visual interactions: Auditory soundscape or visual landscape can influence the perception in a real, multisensory environment (Pheasant et al., 2010; Zhao et al., 2018) that seems to go in one direction, from visual to auditory (Rolls et al., 2022). An interesting McGurke effect (McGurk and MacDonald, 1976) showed an interaction between visual and auditory stimuli, highlighting a higher interaction of otherwise discrete stimulations (Spence and Soto-Faraco, 2010): When a video image of a mouth saying 'g' is played synchronously with playback of the sound 'b', what is perceived is 'd' that generates a sound intermediate in articulation. No study investigating this specific effect is known in animals, but recent research on audiovisual interactions in macaques suggests that monkeys also spontaneously link the auditory and visual components of conspecific calls, preferentially looking at video displays whose mouth shape matches a played call (Fitch, 2018). Interestingly, multisensory interactions are audiovisual speech perception, in which visual speech substantially enhances auditory speech processes (Plass et al., 2020) resulting in more robust effects in audio-visual speech responses than what would have been expected from the summation of the audio and visual speech responses, suggesting that multisensory integration occurs (O'Sullivan et al., 2021).

Auditory-somatosensory interactions: Typical, auditory-somatosensory (AS) humans can multisensory stimulus pairs yield significant reaction time facilitation relative to their unisensory counterparts that exceeded probability (Murray et al., 2005). Bilateral interactions are documented between the auditory and the somatosensory connections (Rolls et al., 2022). Multisensory interactions across space suggest perceptual-cognitive phenomena of sensory-cognitive processing (Foxe et al., 2002). Further, the integration of auditory and somatosensory information in speech processing can be used in a bimodal perceptual task, suggesting somatosensory information on sound categorization can affect adults more than children (Trudeau-Fisette et al., 2019). This results in somatosensory cross-modal reorganization of the auditory cortex in adults with early-stage, mild-moderate ARHL (Cardon and Sharma, 2018). Pairing sound with whisker stimulation modulates tactile responses in both S1 and S2, with the most prominent modulation being robust inhibition in S2 (Zhang et al., 2020). Mutually-inhibitory activation from the S2 circuit can spectrally select tactile versus auditory inputs in mice (Zhang et al., 2020).

Auditory-taste interactions: In contrast to vision and somatosensory interaction, very few examples exist supporting some level of interaction between the auditory and taste systems (Keast and Breslin, 2003; Zampini and Spence, 2012). The strongest is provided through a cross interaction between taste and auditory pitch (Holt-Hansen, 1968; Spence, 2011) that likely provide the impact from the auditory and the insula (Rolls et al., 2022). Most recent evidence suggests an attenuation of taste neophobia induced by taste familiarity and is auditory context-dependent in mice (Grau-Perales and Gallo, 2020). Context dependency involved dopaminergic activity mediated by D1 receptors, which might be responsible for the proper acquisition of safe taste recognition memory (Grau-Perales et al., 2021). In humans, associations between auditory attributes and a number of the commonly agreed basic tastes are now recently added that 'saltiness' (Wang et al., 2021b) among other interactions of taste with auditory interactions (Spence et al., 2021).

Auditory-vestibular interactions: Since rhythm can be biased by passive motion, this suggests that vestibular input may play a vital role in the effect of movement on auditory rhythm processing (Phillips-Silver and Trainor, 2008; Phillips-Silver et al., 2020). Recent evidence suggests that simultaneous auditory–vestibular training facilitates short-term auditory plasticity, producing stronger oscillator connections in the auditory network (Tichko et al., 2021). Most importantly, it can be combined in a vestibular and cochlear prosthesis to restore hearing and balance in patients who have lost both (Phillips et al., 2020). A significant interaction between different sensory modalities during stimulation with a combined vestibular and cochlear prosthesis can help challenge stimulation strategies to simultaneously restore auditory and vestibular function after such an implant (Phillips et al., 2020).

In summary, beyond direct input of A1 now also have higher connections between visual, somatosensory, taste, and vestibular connections to influence and combine with auditory sensation.

6.4. The role of the hippocampus to maintain hearing

Age is associated with a progressive loss of auditory input, starting from the outer hair cells to the inner hair cells followed, with delay, and the loss of sensory neurons (Elliott et al., 2022a). About 1.6 billion people will suffer from hearing impairment (Haile et al., 2021). More recent investigations show an effect of Alzheimer's disease that is affecting hearing progression (Livingston et al., 2020; Michalski and Petit, 2022). Much of this role of Alzheimer's disease was only provided as a novel idea of how the loss of central deletion can formulate a set of suggestions: How can hearing loss affects Alzheimer's and other dementias (Griffiths et al., 2020; Uchida et al., 2022). Griffiths et al. (2020) grouped them into four potential mechanisms based on a common pathology in the cochlea and the brain. The structural and functional features of the auditory brain could play a reciprocal interplay between peripheral and central hearing dysfunction (Johnson et al., 2021). Hearing aids will be helping auditory hearing but currently, there is not sufficient evidence to recommend the use of hearing aids to reduce cognitive decline (Uchida et al., 2022). Moreover, there is no treatment that can cure Alzheimer's disease beyond a moderate decline of progression (van Dyck et al., 2023).

The hippocampus has three branches, CA1-CA3. In addition, the dentate gyrus is expressed in the fibers labeled from the perforant path (green labeled with calretinin; Fig. 52) whereas the granule cells (lilac) are positive for calbindin to differentiate neurons that reach out the nerve fibers (Bottes et al., 2021; Frangou et al., 2021). As the background, granule cells of the hippocampus depend on *Neurod1* and *Lef1* that will never develop (Galceran et al., 2000; Miyata et al., 1999). In contrast to the loss of granule neurons in the absence of *Neurod1* and *Lef1*, we also know that long term production is documented in mice, labeling the newly formed neurons, showing the maturation of adult-generated granule cells that were slower than neonatal-generated granule cells (Overstreet-Wadiche et al., 2006). Note that proliferation is first at the edge of CA3 which will eventually reduce the new granule cells in older mice, showing fewer newly formed neurons (Harris et al., 2021). It appears that the new formation neurons are in certain areas more and show fewer neurons in other areas in older mice and humans (Babcock et al., 2021). Adult hippocampus neurogenesis seems to be regulated by exercise, diet, and social interactions (Uchida et al., 2022).

The hippocampus plays a role in spatial information and memory but it requires establishing how the hippocampus is shaped by sound (Kumar et al., 2021) beyond pleasure sounds (Cheung et al., 2019). A hierarchy to sound can track how the hippocampus affects auditory information (Billig et al., 2022). The evidence suggests a connection from A1, the various belts to reach out perirhinal cortex and the hippocampus (Billig et al., 2022;

The Evolution of the Auditory System as a Blueprint... 183

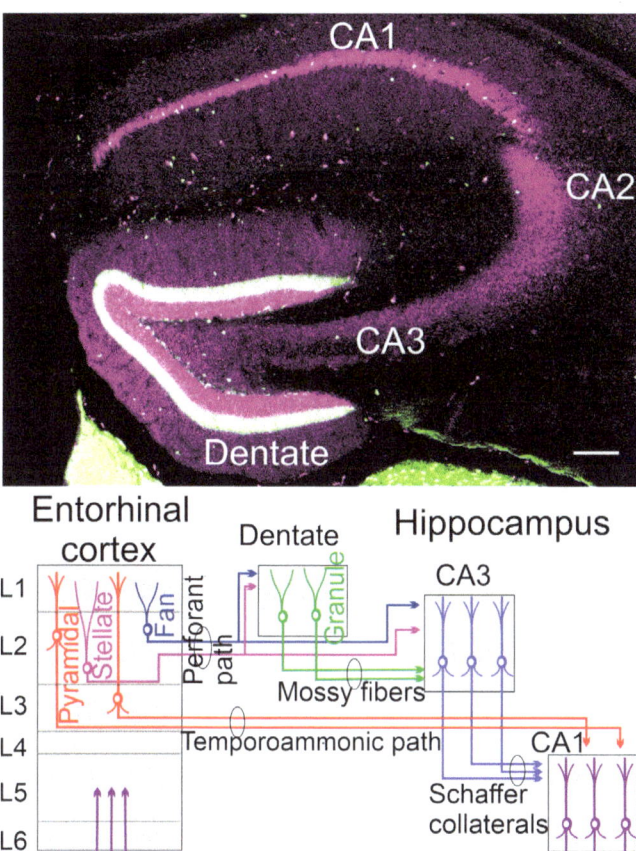

Figure 52: (Top) A coronal section is taken from a 48-day old mouse using calbindin (lilac) and calretinin (green) in a transgenic expression. (Bottom) The dentate gyrus is receiving the perforant pathway from stellate and fan neurons that splits also to innervate the CA3. Mossy fibers extend from granular neurons to innervate the CA3 by pyramidal neurons. From here are the Schaffer collaterals to innervate the CA1 that also receives the input from the temporoammonic path. The process of the fibers returns from CA1 to the entorhinal cortex in L5 by pyramidal neurons (not shown. Unpublished data (top) and modified after (Billig et al., 2022; Igarashi, 2023).

Rolls et al., 2022). Correlating auditory experience can be identified with the extent of broad contributions that ultimately connect hearing loss and dementia, including Alzheimer's disease (Billig et al., 2022).

Pathological changes connect amyloid deposition (*Aβ*), neurofibrillary tangle (*tau*) and combine causing brain atrophy after years (Stopschinski et al., 2021). A very large cohort shows a connection between hearing and in particular temporal lobes (Wang et al., 2022). Poor hearing performance was associated with lower volumes in the amygdala, thalamus, and nucleus accumbens. Hearing impairment was significantly associated with the volume of the hippocampus that disappeared after the Bonferroni correction (Wang et al., 2022). The levels of *tau* were significantly higher in the hearing impairment group which requires further research to establish the mechanism of *tau* and a volume reduction in particular the temporal gyri.

In summary, the hippocampus is involved but is not critical for hearing while *tau* levels show a strong correlation between Alzheimer's and the temporal gyri. Further work is needed to consolidate the hearing connections with the auditory cortex and Alzheimer's disease.

CHAPTER

7

Summary and Conclusion

Making a new auditory system from scratch requires first understanding how the distinction between teleosts and sarcopterygians come about: both auditory systems evolved for hearing but only the sarcopterygians have started with a basilar papilla in sarcopterygians whereas a different sound receiving system evolved in teleosts. A major novel insight is provided by the unique interconnection that could have generated enough pressure during movement to drive the basilar papilla in an outside-in way: from the cochlear aqueduct across the basilar membrane and should have pressure relief via the tympanic membrane across the cartilage. From here, we need to have two independent sound hearing systems: the amphibians and the amniotes. While the cochlear aqueduct places a role for likely all hearing systems that allow the influx from the CSF, whereas amphibians have a complicated perilymphatic system that allows an indirect interaction from the basilar and amphibian papilla across a short recess. Apparently, the perilymphatic system can expand across the brainstem but has no clear opening to allow the CSF which must have some openings in amphibians. Notably, at least two independent tympanic membranes evolved in frogs, sauropsids, and mammals to allow air to drive the tympanic membrane with stapes to insert into the oval window. From the oval window, a sound system can drive across the perilymphatic space to exit via the round window. Having deciphered the tetrapod, sound production requires the next steps: developing hair cells that link up, via perilymph connections, to provide unique sensory neurons to reach out the auditory nuclei in most tetrapods, except specific amphibians that lose auditory input.

Auditory hair cells evolved in sarcopterygians of the basilar papilla about 400 million years ago and are lost in lungfish and several amphibians. Within amniotes, the range can be small (about 60 hair cells) to very large in certain sauropsids and mammals with several thousand hair cells. In contrast to limited morphology changes in sauropsids contrast with mammalian hearing that is consisting of a single row of inner hair cells that is separated by the pillar cells to have three rows of outer hair cells. The innervation will be innervated by a separate set of fibers in most amniotes that overlaps with the amphibian sensory neurons. Spiral ganglion neurons are a unique innervation in the mammalian cochlea that shows about 95% that innervates the inner hair cells whereas about 5% reach out to the outer hair cells. No such distinct differences exists in sauropsids that innervate the various length of hair cells.

Similarity exists for the cochlear nuclei in amniotes but also shows a unique input of the length of cochlear nuclei that varies between four rhombomeres in mammals (r2-r5) whereas in sauropsids it extends between r2-r8 that could equally extend caudally also in frogs. This is somewhat similar in cochlear nuclei organization between the amniotes but also shows more distinct nuclei that differ in various outputs to reach out the superior olivary, lateral lemniscus, and the inferior colliculi. Overall, a somewhat central projection beyond the cochlear nuclei is known in frogs that show less distinct cochlear nuclei. From here we see a common output to go from the inferior colliculi to the medial geniculate body that reaches out in frogs, birds, and mammals which have a tonotopic organization across all tetrapods. In mammals and birds, we can see an increasing integration of input from all cochlear output to provide sound processing like interaural time, level, and phase differences for ITD, ILD, and IPD that integrates into the inferior colliculi, supported by reciprocal input from the auditory cortex and the medial geniculate body.

The auditory cortex receives a tonotopic input in birds and mammals but does not have a tonotopic input in the frog's cortex. Within mammals, we can have a major auditory cortex surrounded by belts that receive contact from the medial geniculate body to have a reciprocal connection. From here we can have an output to reach out the Broca's area in humans that has an equivalent bird output. Singing is common among various mammals, birds, frogs, and specific teleosts, but a combination of singing and speaking is a unique feature in humans. We have a long time of one year or more as an infant before speaking and singing starts during which time connections will be established for sound processing. For that language and singing early development, we have no data beyond already speaking humans. We also have a brought interaction for other sensory inputs that coalesce into combined interaction from somatosensory, vestibular, taste, and olfaction. As the sequence of the auditory cortex will eventually reach out to the hippocampus that likely plays no major role for hearing but could have some auditory processing. Understanding the role of transcription factors in gene regulation is crucial for understanding hearing development and evolution. Transcription factors are proteins that bind to specific DNA sequences to control the expression of genes. In the case of hearing, transcription factors play a critical role in the development of various cell types and structures within the auditory system.

The bHLH (basic helix-loop-helix) family of transcription factors has been shown to have distinct functions in different regions of the developing auditory system. For example, in the brainstem, *Atoh1* is needed for the development of hair cells, which are the sensory receptors in the inner ear that detect sound. *Atoh1* also works with *Ptf1a* to develop the cochlear nuclei, which are groups of neurons in the brainstem that receive and process auditory information from the cochlea.

In contrast, spiral ganglion neurons, which transmit auditory information from the hair cells to the brainstem, depend on a different set of transcription factors. These include *Neurog1* followed by *Neurod1*, which are involved in the development and differentiation of these neurons.

The inferior colliculi, which are part of the midbrain auditory pathway, rely on a novel set of genes for their development. Later in development, they express *Neurod1*, *Neurog2*, and *Ascl1* across the medial geniculate body and the auditory cortex.

Overall, the regulation of gene expression by transcription factors is essential for the proper development and function of the auditory system, and understanding their roles is critical for developing treatments for hearing disorders.

Overall, from *Latimeria* to humans, we have an increasing sound input to require processing the various tetrapods. What makes it unique in humans is the integrating from the auditory cortex that provides a lengthy learning from becoming able to speak, something we will eventually lose as hearing and appropriate sound input in older humans, that makes it so different compared to birds that can replace hair cells easily. Understanding why our system follows such a unique developmental path will help restart hearing in older people. Perhaps it is the high genetic load in humans that can correlate with the loss of hearing. We are at the beginning of future research to understand and help to hear in very young infants as well as older people.

References

Agoston, Z., Li, N., Haslinger, A., Wizenmann, A. and Schulte, D. (2012). Genetic and physical interaction of Meis2, Pax3 and Pax7 during dorsal midbrain development. BMC Dev Biol *12*, 10.

Altman, J. and Bayer, S.A. (1989). Development of the rat thalamus: V. The posterior lobule of the thalamic neuroepithelium and the time and site of origin and settling pattern of neurons of the medial geniculate body. Journal of Comparative Neurology *284*, 567–580.

Amemiya, C.T., Alföldi, J., Lee, A.P., Fan, S., Philippe, H., MacCallum, I., Braasch, I., Manousaki, T., Schneider, I. and Rohner, N. (2013). The African coelacanth genome provides insights into tetrapod evolution. Nature *496*, 311–316.

Arnold, S.J., Huang, G., Cheung, A.F.P., Era, T., Nishikawa, S.I., Bikoff, E.K., Molnár, Z., Robertson, E.J. and Groszer, M. (2008). The T-box transcription factor Eomes/Tbr2 regulates neurogenesis in the cortical subventricular zone. Genes & Development *22(18)*, 2479–2484.

Atturo, F., Schart-Morén, N., Larsson, S., Rask-Andersen, H. and Li, H. (2018). The human cochlear aqueduct and accessory canals: A micro-CT analysis using a 3D reconstruction paradigm. Otology & Neurotology *39*, e429-e435.

Babcock, K.R., Page, J.S., Fallon, J.R. and Webb, A.E. (2021). Adult hippocampal neurogenesis in aging and Alzheimer's disease. Stem Cell Reports *16*, 681–693.

Baker, C.V. (2019). The development and evolution of lateral line electroreceptors: Insights from comparative molecular approaches. *In:* Electroreception: Fundamental Insights from Comparative Approaches (Springer), pp. 25–62.

Barkat, T.R., Polley, D.B. and Hensch, T.K. (2011). A critical period for auditory thalamocortical connectivity. Nature Neuroscience *14*, 1189–1194.

Bass, A.H., Chagnaud, B.P. and Feng, N.Y. (2015). Comparative Neurobiology of Sound Production in Fishes. *In:* Sound Communication in Fishes. Animal Signals and Communication, vol 4. F. Ladich ed. (Springer, Vienna), pp. 35–75.

Bemis, W.E. and Northcutt, R.G. (1991). Innervation of the basicranial muscle of Latimeria chalumnae. *In:* The Biology of *Latimeria chalumnae* and Evolution of Coelacanths (Springer), pp. 147–158.

Bernstein, P. (2003). The ear region of *Latimeria chalumnae:* Functional and evolutionary implications. Zoology *106*, 233–242.

Billig, A.J., Lad, M., Sedley, W. and Griffiths, T.D. (2022). The hearing hippocampus. Progress in Neurobiology *218*, 102326.

Birinyi, A., Straka, H., Matesz, C. and Dieringer, N. (2001). Location of dye-coupled second order and of efferent vestibular neurons labeled from individual semicircular canal or otolith organs in the frog. Brain Research *921*, 44–59.

Bjerring, H.C. (1985). Facts and thoughts on piscine phylogeny. *In:* Evolutionary Biology of Primitive Fishes (Springer), pp. 31–57.

Bleckmann, H., Niemann, U. and Fritzsch, B. (1991). Peripheral and central aspects of the acoustic and lateral line system of a bottom dwelling catfish, *Ancistrus* sp. Journal of Comparative Neurology *314*, 452–466.

Boltzmann, L. (1887). XXXVI. On the assumptions necessary for the theoretical proof of Avogadro's law. The London, Edinburgh and Dublin, Philosophical Magazine and Journal of Science *23*, 305–333.

Bottes, S., Jaeger, B.N., Pilz, G.A., Jörg, D.J., Cole, J.D., Kruse, M., Harris, L., Korobeynyk, V.I., Mallona, I., Helmchen, F. *et al.* (2021). Long-term self-renewing stem cells in the adult mouse hippocampus identified by intravital imaging. Nat Neurosci *24*, 225–233.

Budinger, E. (2020). Primary auditory cortex and the thalamo-cortico-thalamic circuitry I. Anatomy. *In:* The Senses: A Comprehensive Reference (Second Edition), B. Fritzsch, ed. (Oxford: Elsevier), pp. 623–656.

Budinger, E. and Kanold, P.O. (2018). Auditory cortex circuits. *In:* The Mammalian Auditory Pathways (Springer), pp. 199–233.

Bullock, T.H. and Hopkins, C.D. (2005). Explaining Electroreception (Springer).

Burkhardt, P. (2022). Ctenophores and the evolutionary origin(s) of neurons. Trends Neurosci *45*, 878–880.

Cai, X., Kardon, A.P., Snyder, L.M., Kuzirian, M.S., Minestro, S., de Souza, L., Rubio, M.E., Maricich, S.M. and Ross, S.E. (2016). BHLHb5: flpo allele uncovers a requirement for BHLHb5 for the development of the dorsal cochlear nucleus. Developmental Biology *414*, 149–160.

Capshaw, G., Christensen-Dalsgaard, J. and Carr, C.E. (2022). Hearing without a tympanic ear. Journal of Experimental Biology *225*, jeb244130.

Capshaw, G., Soares, D., Christensen-Dalsgaard, J. and Carr, C. (2020). Seismic sensitivity and bone conduction mechanisms enable extratympanic hearing in salamanders. Journal of Experimental Biology *223*, jeb236489.

Carbajal, G.V. and Malmierca, M.S. (2020). Novelty processing in the auditory system: Detection, adaptation or expectation? *In:* The Senses: A Comprehensive Reference (Second Edition), B. Fritzsch, ed. (Oxford: Elsevier), pp. 749–776.

Cardon, G. and Sharma, A. (2018). Somatosensory cross-modal reorganization in adults with age-related, early-stage hearing loss. Frontiers in Human Neuroscience *12*, 172.

Carr, C.E. (2020). Evolution of Central Pathways. *In:* The Senses, B. Fritzsch, ed. (Elsevier), pp. 354–376.

Chagnaud, B.P., Engelmann, J., Fritzsch, B., Glover, J.C. and Straka, H. (2017). Sensing external and self-motion with hair cells: A comparison of the lateral line and vestibular systems from a developmental and evolutionary perspective. Brain, Behavior and Evolution *90*, 98–116.

Chambers, A.R., Resnik, J., Yuan, Y., Whitton, J.P., Edge, A.S., Liberman, M.C. and Polley, D.B. (2016). Central gain restores auditory processing following near-complete cochlear denervation. Neuron *89*, 867–879.

Chapuis, L. and Collin, S.P. (2022). The auditory system of cartilaginous fishes. Reviews in Fish Biology and Fisheries, 1–34.

Cheung, V.K., Harrison, P.M., Meyer, L., Pearce, M.T., Haynes, J.-D. and Koelsch, S. (2019). Uncertainty and surprise jointly predict musical pleasure and amygdala, hippocampus, and auditory cortex activity. Current Biology *29*, 4084–4092. e4084.

Chizhikov, V.V., Iskusnykh, I.Y., Fattakhov, N. and Fritzsch, B. (2021). Lmx1a and Lmx1b are redundantly required for the development of multiple components of the mammalian auditory system. Neuroscience *452*, 247–264.

Christensen, C.B., Christensen-Dalsgaard, J. and Madsen, P.T. (2015a). Hearing of the African lungfish (*Protopterus annectens*) suggests underwater pressure detection and rudimentary aerial hearing in early tetrapods. Journal of Experimental Biology *218*, 381–387.

Christensen, C.B., Lauridsen, H., Christensen-Dalsgaard, J., Pedersen, M. and Madsen, P.T. (2015b). Better than fish on land? Hearing across metamorphosis in salamanders. Proceedings of the Royal Society B: Biological Sciences *282*, 20141943.

Chumak, T., Tothova, D., Filova, I., Bures, Z., Popelar, J., Pavlinkova, G. and Syka, J. (2021). Overexpression of Isl1 under the Pax2 promoter, leads to impaired sound processing and increased inhibition in the inferior colliculus. International Journal of Molecular Sciences *22*, 4507.

Clack, J.A. and Ahlberg, P.E. (2016). Sarcopterygians: From lobe-finned fishes to the tetrapod stem group. *In:* Evolution of the Vertebrate Ear (Springer), pp. 51–70.

Corwin, J.T. (1981). Postembryonic production and aging of inner ear hair cells in sharks. Journal of Comparative Neurology *201*, 541–553.

Cox, C., Bergmann, C., Fowler, E., Keren-Portnoy, T., Roepstorff, A., Bryant, G. and Fusaroli, R. (2023). A systematic review and Bayesian meta-analysis of the acoustic features of infant-directed speech. Nat Hum Behav *7*, 114–133.

Daffern, H., Keren-Portnoy, T., DePaolis, R.A. and Brown, K.I. (2020). BabblePlay: An app for infants, controlled by infants, to improve early language outcomes. Appl Acoust *162*, 107183.

de Burlet, H. (1934). Vergleichende Anatomie des stato-akustischen Organs. Handbuch der vergleichenden Anatomie der Wirbeltiere *2*, 1293–1432.

Dean, J. and Claas, B. (2020). Hydrodynamic sensing by the African clawed frog, Xenopus laevis. *In:* The Senses: A Comprehensive Reference (Second Edition), B. Fritzsch, ed. (Oxford: Elsevier), pp. 185–214.

Dempsey, C., Abbott, L.F. and Sawtell, N.B. (2019). Generalization of learned responses in the mormyrid electrosensory lobe. Elife *8*, e44032.

Dennis, D.J., Han, S. and Schuurmans, C. (2019). bHLH transcription factors in neural development, disease, and reprogramming. Brain Research *1705*, 48–65.

Di Bonito, M. and Studer, M. (2017). Cellular and molecular underpinnings of neuronal assembly in the central auditory system during mouse development. Frontiers in Neural Circuits *11*, 18.

Di Bonito, M., Studer, M. and Puelles, L. (2017). Nuclear derivatives and axonal projections originating from rhombomere 4 in the mouse hindbrain. Brain Structure and Function 222, 3509–3542.

Diek, D., Smidt, M.P. and Mesman, S. (2022). Molecular organization and patterning of the medulla oblongata in health and disease. International Journal of Molecular Sciences 23, 9260.

Dobzhansky, T. (2013). Nothing in biology makes sense except in the light of evolution. The American Biology Teacher 75, 87–91.

Driscoll, M.E. and Tadi, P. (2022). Neuroanatomy, Inferior Colliculus. StatPearls Publishing, Treasure Island (FL).

Dutel, H., Galland, M., Tafforeau, P., Long, J.A., Fagan, M.J., Janvier, P., Herrel, A., Santin, M.D., Clément, G. and Herbin, M. (2019). Neurocranial development of the coelacanth and the evolution of the sarcopterygian head. Nature 569, 556–559.

Dvorakova, M., Macova, I., Bohuslavova, R., Anderova, M., Fritzsch, B. and Pavlinkova, G. (2020). Early ear neuronal development, but not olfactory or lens development, can proceed without SOX2. Developmental biology 457, 43–56.

Eggenschwiler, J.T., Espinoza, E. and Anderson, K.V. (2001). Rab23 is an essential negative regulator of the mouse Sonic hedgehog signalling pathway. Nature 412, 194–198.

Ehret, G., Tautz, J., Schmitz, B. and Narins, P. (1990). Hearing through the lungs: Lung-eardrum transmission of sound in the frog *Eleutherodactylus coqui*. Naturwissenschaften 77, 192–194.

Elemans, C.P., Rasmussen, J.H., Herbst, C.T., Düring, D.N., Zollinger, S.A., Brumm, H., Srivastava, K., Svane, N., Ding, M. and Larsen, O.N. (2015). Universal mechanisms of sound production and control in birds and mammals. Nature Communications 6, 1–13.

Elliott, E.J. and Smart, D.R. (2014). The assessment and management of inner ear barotrauma in divers and recommendations for returning to diving. Diving Hyperb Med 44, 208–222.

Elliott, K.L., Kersigo, J., Pan, N., Jahan, I. and Fritzsch, B. (2017). Spiral ganglion neuron projection development to the hindbrain in mice lacking peripheral and/or central target differentiation. Frontiers in Neural Circuits 11, 25.

Elliott, K.L., Fritzsch, B. and Duncan, J.S. (2018). Evolutionary and developmental biology provide insights into the regeneration of organ of Corti hair cells. Frontiers in Cellular Neuroscience 12, 252.

Elliott, K.L. and Fritzsch, B. (2020). Evolution and development of lateral line and electroreception: An integrated perception of neurons, hair cells and brainstem nuclei. *In:* The Senses (Elsevier Inc), pp. 95–115.

Elliott, K.L., Kersigo, J., Lee, J.H., Jahan, I., Pavlinkova, G., Fritzsch, B. and Yamoah, E.N. (2021a). Developmental changes in peripherin-eGFP expression in spiral ganglion neurons. Frontiers in Cellular Neuroscience 15, 678113.

Elliott, K.L., Kersigo, J., Lee, J.H., Yamoah, E.N. and Fritzsch, B. (2021b). Sustained loss of Bdnf affects peripheral but not central vestibular targets. Frontiers in Neurology 12, 768456–768456.

Elliott, K.L., Pavlínková, G., Chizhikov, V.V., Yamoah, E.N. and Fritzsch, B. (2021c). Development in the mammalian auditory system depends on transcription factors. International Journal of Molecular Sciences 22, 4189.

Elliott, K.L., Fritzsch, B., Yamoah, E.N. and Zine, A. (2022). Age-related hearing loss: Sensory and neural etiology and their interdependence. Frontiers in Aging Neuroscience 14, 814526.

Elliott, K.L. and Straka, H. (2022). Assembly and functional organization of the vestibular system. In: Evolution of Neurosensory Cells and Systems (CRC Press), pp. 135–174.

Elliott, K.L., Iskusnykh, I.Y., Chizhikov, V.V. and Fritzsch, B. (2023). Ptf1a expression is necessary for correct targeting of spiral ganglion neurons within the cochlear nuclei. Neuroscience Letters 806, 137244, DOI: 10.1016/j.neulet.2023.137244

Enard, W. (2011). FOXP2 and the role of cortico-basal ganglia circuits in speech and language evolution. Current Opinion in Neurobiology 21, 415–424.

Engelmann, J. and Fritzsch, B. (2022). Lateral line input to 'almost' all vertebrates shares a common organization with different distinct connections. In: Evolution of Neurosensory Cells and Systems (CRC Press), pp. 201–222.

Engelmann, J., Hanke, W., Mogdans, J. and Bleckmann, H. (2000). Neurobiology: Hydrodynamic stimuli and the fish lateral line. Nature 408, 51–52.

Felmy, F. and Meyer, E.M.M. (2020). Lateral Lemniscus. In: The Senses: A Comprehensive Reference (Second Edition), B. Fritzsch, ed. (Oxford: Elsevier), pp. 556–565.

Feng, A.S. and Lin, W.Y. (1991). Differential innervation patterns of three divisions of frog auditory midbrain (torus semicircularis). J Comp Neurol 306, 613–630.

Feng, A.S. and Schellart, N.A.M. (1999). Central Auditory Processing in Fish and Amphibians. Comparative Hearing: Fish and Amphibians 218–268.

Filova, I., Dvorakova, M., Bohuslavova, R., Pavlinek, A., Elliott, K.L., Vochyanova, S., Fritzsch, B. and Pavlinkova, G. (2020). Combined Atoh1 and Neurod1 deletion reveals autonomous growth of auditory nerve fibers. Molecular Neurobiology 57, 5307–5323.

Filova, I., Bohuslavova, R., Tavakoli, M., Yamoah, E.N., Fritzsch, B. and Pavlinkova, G. (2022a). Early deletion of Neurod1 alters neuronal lineage potential and diminishes neurogenesis in the inner ear. Frontiers in Cell and Developmental Biology 10, 845461–845461.

Filova, I., Pysanenko, K., Tavakoli, M., Vochyanova, S., Dvorakova, M., Bohuslavova, R., Smolik, O., Fabriciova, V., Hrabalova, P., Benesova, S. et al. (2022b). ISL1 is necessary for auditory neuron development and contributes toward tonotopic organization. Proc Natl Acad Sci USA 119, e2207433119.

Fitch, W.T. (2018). The biology and evolution of speech: A comparative analysis. Annual Review of Linguistics 4, 255–279.

Forey, P.L. (1998). History of the Coelacanth Fishes. London: Chapman & Hall.

Foxe, J.J., Wylie, G.R., Martinez, A., Schroeder, C.E., Javitt, D.C., Guilfoyle, D., Ritter, W. and Murray, M.M. (2002). Auditory-somatosensory multisensory processing in auditory association cortex: An fMRI study. Journal of Neurophysiology 88, 540–543.

Frangou, S., Modabbernia, A., Williams, S.C.R., Papachristou, E., Doucet, G.E., Agartz, I., Aghajani, M., Akudjedu, T.N., Albajes-Eizagirre, A., Alnaes, D. et al. (2021). Cortical thickness across the lifespan: Data from 17,075 healthy individuals aged 3-90 years. Hum Brain Mapp 43, 431–451.

Franklin, R.E. and Gosling, R.G. (1953). Molecular configuration in sodium thymonucleate. Nature 171, 740–741.

Fritzsch, B. (1987). Inner ear of the coelacanth fish *Latimeria* has tetrapod affinities. Nature *327*, 153–154.

Fritzsch, B. (1990). The evolution of metamorphosis in amphibians. Journal of Neurobiology *21*, 1011–1021.

Fritzsch, B. (1992). The water-to-land transition: Evolution of the tetrapod basilar papilla, middle ear, and auditory nuclei. *In:* The Evolutionary Biology of Hearing, D. Webster, R.R. Fay and A.N. Popper, eds. (Berlin: Springer), pp. 351–375.

Fritzsch, B. (1998). Evolution of the vestibulo-ocular system. Otolaryngology—Head and Neck Surgery *119*, 182–192.

Fritzsch, B. (2003). The ear of *Latimeria chalumnae* revisited. ZOOLOGY-JENA *106*, 243–248.

Fritzsch, B. (2021). An integrated perspective of evolution and development: From genes to function to ear, lateral line and electroreception. Diversity *13*, 364.

Fritzsch, B. and Wake, M. (1988). The inner ear of gymnophione amphibians and its nerve supply: A comparative study of regressive events in a complex sensory system (Amphibia, Gymnophiona). Zoomorphology *108*, 201–217.

Fritzsch, B. and Neary, T. (1998). The octavolateralis system of mechanosensory and electrosensory organs. Amphibian Biology *3*, 878–922.

Fritzsch, B. and Beisel, K. (2004). Keeping sensory cells and evolving neurons to connect them to the brain: Molecular conservation and novelties in vertebrate ear development. Brain, Behavior and Evolution *64*, 182–197.

Fritzsch, B., Pauley, S. and Beisel, K.W. (2006). Cells, molecules and morphogenesis: The making of the vertebrate ear. Brain Research *1091*, 151–171.

Fritzsch, B., Dillard, M., Lavado, A., Harvey, N.L. and Jahan, I. (2010). Canal cristae growth and fiber extension to the outer hair cells of the mouse ear require Prox1 activity. PloS One *5*, e9377.

Fritzsch, B., Pan, N., Jahan, I., Duncan, J.S., Kopecky, B.J., Elliott, K.L., Kersigo, J. and Yang, T. (2013). Evolution and development of the tetrapod auditory system: An organ of Corti-centric perspective. Evolution & Development *15*, 63–79.

Fritzsch, B. and Straka, H. (2014). Evolution of vertebrate mechanosensory hair cells and inner ears: Toward identifying stimuli that select mutation driven altered morphologies. Journal of Comparative Physiology A *200*, 5–18.

Fritzsch, B., Pan, N., Jahan, I. and Elliott, K.L. (2015). Inner ear development: Building a spiral ganglion and an organ of Corti out of unspecified ectoderm. Cell and Tissue Research *361*, 7–24.

Fritzsch, B. and Elliott, K.L. (2017). Gene, cell, and organ multiplication drives inner ear evolution. Developmental Biology *431*, 3–15.

Fritzsch, B., Elliott, K.L. and Pavlinkova, G. (2019). Primary sensory map formations reflect unique needs and molecular cues specific to each sensory system. F1000Research *8*.

Fritzsch, B., Erives, A., Eberl, D.F. and Yamoah, E.N. (2020). Genetics of mechanoreceptor evolution and development. *In:* The Senses, B. Fritzsch, ed. (Elsevier Inc), pp. 277–301.

Fritzsch, B., Elliott, K.L. and Yamoah, E.N. (2022). Neurosensory development of the four brainstem-projecting sensory systems and their integration in the telencephalon. Frontiers in Neural Circuits *16*, 913480.

Fritzsch, B. and Martin, P.R. (2022). Vision and retina evolution: How to develop a retina. IBRO Neuroscience Reports *12*, 240–248.

Fritzsch, B., Kersigo, J., Rejent, K., Gherman, W., Frank, P.W., Giovannucci, D.R. and Maklad, A. (2023a). Hair cell morphological patterns and polarity organization in the sea lamprey vestibular cristae. Anat Rec (Hoboken).

Fritzsch, B., Schultze, H.P. and Elliott, K.L. (2023b). The evolution of the various structures required for hearing in Latimeria and tetrapods. IBRO Neuroscience Reports *14*, 325–341.

Fujiyama, T., Yamada, M., Terao, M., Terashima, T., Hioki, H., Inoue, Y.U., Inoue, T., Masuyama, N., Obata, K., Yanagawa, Y., Kawaguchi, Y., Nabeshima, Y. and Hoshino, M. (2009). Inhibitory and excitatory subtypes of cochlear nucleus neurons are defined by distinct bHLH transcription factors, Ptf1a and Atoh1. Development *136(12)*, 2049–2058.

Galceran, J., Miyashita-Lin, E.M., Devaney, E., Rubenstein, J. and Grosschedl, R. (2000). Hippocampus development and generation of dentate gyrus granule cells is regulated by LEF1. Development *127*, 469–482.

García-Añoveros, J., Clancy, J.C., Foo, C.Z., García-Gómez, I., Zhou, Y., Homma, K., Cheatham, M.A. and Duggan, A. (2022). Tbx2 is a master regulator of inner versus outer hair cell differentiation. Nature *605*, 298–303.

Glover, J.C., Elliott, K.L., Erives, A., Chizhikov, V.V. and Fritzsch, B. (2018). Wilhelm His' lasting insights into hindbrain and cranial ganglia development and evolution. Developmental Biology *444*, S14-S24.

Glover, J.C. and Fritzsch, B. (2022). Molecular mechanisms governing development of the hindbrain choroid plexus and auditory projection: A validation of the seminal observations of Wilhelm His. IBRO Neuroscience Reports *13*, 306–313.

Goodrich, L. and Kanold, P. (2020). Functional circuit development in the auditory system. *In:* Neural Circuit and Cognitive Development (Elsevier), pp. 27–55.

Grau-Perales, A. and Gallo, M. (2020). The auditory context-dependent attenuation of taste neophobia depends on D1 dopamine receptor activity in mice. Behavioural Brain Research *391*, 112687.

Grau-Perales, A.B., Gámiz, F. and Gallo, M. (2021). Effect of hippocampal 6-OHDA lesions on the contextual modulation of taste recognition memory. Behavioural Brain Research *409*, 113320.

Griffiths, T.D., Lad, M., Kumar, S., Holmes, E., McMurray, B., Maguire, E.A., Billig, A.J. and Sedley, W. (2020). How can hearing loss cause dementia? Neuron *108*, 401–412.

Grothe, B. (2020). The auditory system function – An integrative perspective. *In:* The Senses: A Comprehensive Reference (Second Edition), B. Fritzsch, ed. (Oxford: Elsevier), pp. 1–17.

Grothe, B., Carr, C.E., Casseday, J.H., Fritzsch, B. and Köppl, C. (2004). The evolution of central pathways and their neural processing patterns. *In:* Evolution of the Vertebrate Auditory System (Springer), pp. 289–359.

Gurung, B. and Fritzsch, B. (2004). Time course of embryonic midbrain and thalamic auditory connection development in mice as revealed by carbocyanine dye tracing. J Comp Neurol *479*, 309–327.

Haeckel, E. (1876). Die Perigenesis der Plastidule oder die Wellenzeugung der Lebenstheilchen: Ein Versuch zur mechanischen Erklärung der elementaren Entwickelungs-Vorgänge (De Gruyter, Incorporated).

Haile, L.M., Kamenov, K., Briant, P.S., Orji, A.U., Steinmetz, J.D., Abdoli, A., Abdollahi, M., Abu-Gharbieh, E., Afshin, A. and Ahmed, H. (2021). Hearing loss prevalence and years lived with disability, 1990–2019: Findings from the Global Burden of Disease Study 2019. The Lancet *397*, 996–1009.

Harper, T. and Rougier, G.W. (2019). Petrosal morphology and cochlear function in Mesozoic stem therians. PLoS One *14*, e0209457.

Harris, L., Rigo, P., Stiehl, T., Gaber, Z.B., Austin, S.H., del Mar Masdeu, M., Edwards, A., Urbán, N., Marciniak-Czochra, A. and Guillemot, F. (2021). Coordinated changes in cellular behavior ensure the lifelong maintenance of the hippocampal stem cell population. Cell Stem Cell *28*, 863-876. e866.

Henke, R.M., Meredith, D.M., Borromeo, M.D., Savage, T.K. and Johnson, J.E. (2009). Ascl1 and Neurog2 form novel complexes and regulate Delta-like3 (Dll3) expression in the neural tube. Dev Biol *328*, 529-540.

Herculano-Houzel, S. (2016). The Human Advantage: A New Understanding of How Our Brain Became Remarkable. MIT Press.

Hernandez-Miranda, L.R., Müller, T. and Birchmeier, C. (2017). The dorsal spinal cord and hindbrain: From developmental mechanisms to functional circuits. Developmental Biology *432*, 34–42.

Herrick, C.J. (1948). The brain of the tiger salamander, Ambystoma tigrinum. University of Chicago Press. https://doi.org/10.5962/bhl.title.6375

Hevner, R.F., Shi, L., Justice, N., Hsueh, Y.-P., Sheng, M., Smiga, S., Bulfone, A., Goffinet, A.M., Campagnoni, A.T. and Rubenstein, J.L. (2001). Tbr1 regulates differentiation of the preplate and layer 6. Neuron *29*, 353–366.

Holt-Hansen, K. (1968). Taste and pitch. Perceptual and Motor Skills *27*, 59–68.

Horng, S., Kreiman, G., Ellsworth, C., Page, D., Blank, M., Millen, K. and Sur, M. (2009). Differential gene expression in the developing lateral geniculate nucleus and medial geniculate nucleus reveals novel roles for Zic4 and Foxp2 in visual and auditory pathway development. Journal of Neuroscience *29*, 13672–13683.

Hornibrook, J. (2018). The postural and cognitive disabilites of chronic Perilymph Fistula (PLF) after mild head trauma. Annals of Clinical Case Reports-Otolaryngology *3*, 1514.

Hoshino, N., Altarshan, Y., Alzein, A., Fernando, A.M., Nguyen, H.T., Majewski, E.F., Chen, V.C., Rochlin, M.W. and Yu, W.M. (2021). Ephrin-A3 is required for tonotopic map precision and auditory functions in the mouse auditory brainstem. J Comp Neurol *529*, 3633-3654.

Igarashi, K.M. (2023). Entorhinal cortex dysfunction in Alzheimer's disease. Trends Neurosci *46*, 124–136.

Iskusnykh, I.Y., Steshina, E.Y. and Chizhikov, V.V. (2016). Loss of Ptf1a leads to a widespread cell-fate misspecification in the brainstem, affecting the development of somatosensory and viscerosensory nuclei. Journal of Neuroscience *36*, 2691–2710.

Ito, T. and Malmierca, M.S. (2018). Neurons, connections, and microcircuits of the inferior colliculus. *In:* The Mammalian Auditory Pathways (Springer), pp. 127–167.

Iulianella, A., Wingate, R.J., Moens, C.B. and Capaldo, E. (2019). The generation of granule cells during the development and evolution of the cerebellum. Developmental Dynamics *248*, 506–513.

Jahan, I., Kersigo, J., Pan, N. and Fritzsch, B. (2010). Neurod1 regulates survival and formation of connections in mouse ear and brain. Cell and Tissue Research *341*, 95–110.

Jahan, I., Pan, N., Elliott, K.L. and Fritzsch, B. (2015a). The quest for restoring hearing: Understanding ear development more completely. Bioessays *37*, 1016–1027.

Jahan, I., Pan, N., Kersigo, J. and Fritzsch, B. (2015b). Neurog1 can partially substitute for Atoh1 function in hair cell differentiation and maintenance during organ of Corti development. Development *142*, 2810–2821.

Jahan, I., Kersigo, J., Elliott, K.L. and Fritzsch, B. (2021). Smoothened overexpression causes trochlear motoneurons to reroute and innervate ipsilateral eyes. Cell and Tissue Research *384*, 59–72.

Jaslow, A., Hetherington, T. and Lombard, R. (1988). Structure and function of the amphibian middle ear. *In:* The Evolution of the Amphibian Auditory System, B. Fritzsch, M.J. Ryan, W. Wilczynski, T. Hetherington and W. Walkowiak, eds. (New York: Wylie), pp. 69–91.

Jeong, H., Clark, S., Goehring, A., Dehghani-Ghahnaviyeh, S., Rasouli, A., Tajkhorshid, E. and Gouaux, E. (2022). Structures of the TMC-1 complex illuminate mechanosensory transduction. Nature, 1–8.

Ji, Y.R., Tona, Y., Wafa, T., Christman, M.E., Tourney, E.D., Jiang, T., Ohta, S., Cheng, H., Fitzgerald, T. and Fritzsch, B. (2022). Function of bidirectional sensitivity in the otolith organs established by transcription factor Emx2. Nature Communications *13*, 1–14.

Johnson, J.C., Marshall, C.R., Weil, R.S., Bamiou, D.-E., Hardy, C.J. and Warren, J.D. (2021). Hearing and dementia: From ears to brain. Brain *144*, 391–401.

Jung, S.N., Lucks, V., Elliott, K.L. and Fritzsch, B. (2022). Electroreception depends on hair cell-derived senses in some vertebrates. *In:* Evolution of Neurosensory Cells and Systems (CRC Press), pp. 223–254.

Kaiser, M., Wojahn, I., Rudat, C., Lüdtke, T.H., Christoffels, V.M., Moon, A., Kispert, A. and Trowe, M.O. (2021). Regulation of otocyst patterning by Tbx2 and Tbx3 is required for inner ear morphogenesis in the mouse. Development *148*, dev195651

Kandler, K., Lee, J. and Pecka, M. (2020). The Superior Olivary Complex. *In:* The Senses: A Comprehensive Reference (Second Edition), B. Fritzsch, ed. (Oxford: Elsevier), pp. 533–555.

Kanold, P.O. and Luhmann, H.J. (2010). The subplate and early cortical circuits. Annual Review of Neuroscience *33*, 23–48.

Kanold, P.O., Nelken, I. and Polley, D.B. (2014). Local versus global scales of organization in auditory cortex. Trends in Neurosciences *37*, 502–510.

Karmakar, K., Narita, Y., Fadok, J., Ducret, S., Loche, A., Kitazawa, T., Genoud, C., Di Meglio, T., Thierry, R. and Bacelo, J. (2017). Hox2 genes are required for tonotopic map precision and sound discrimination in the mouse auditory brainstem. Cell Reports *18*, 185–197.

Keast, R.S. and Breslin, P.A. (2003). An overview of binary taste-taste interactions. Food Quality and Preference *14*, 111–124.

Keeling, P., Leander, B. and Simpson, A. (2009). Eukaryotes: Eukaryota, organisms with nucleated cells. The Tree of Life Project, http://tolweb org/Eukaryotes/3/200910 *28*.

Kelley, D.B., Ballagh, I.H., Barkan, C.L., Bendesky, A., Elliott, T.M., Evans, B.J., Hall, I.C., Kwon, Y.M., Kwong-Brown, U. and Leininger, E.C. (2020). Generation, coordination, and evolution of neural circuits for vocal communication. Journal of Neuroscience 40, 22–36.

Kempf, J.M., Knelles, K., Hersbach, B.A., Petrik, D., Riedemann, T., Bednářová, V., Janjić, A., Simon-Ebert, T., Enard, W., Smialowski, P. *et al.* (2021). Heterogeneity of neurons reprogrammed from spinal cord astrocytes by the proneural factors Ascl1 and Neurogenin2. Cell Reports 36, 109409.

Keren-Portnoy, T., Majorano, M. and Vihman, M.M. (2009). From phonetics to phonology: The emergence of first words in Italian. Journal of Child Language 36, 235–267.

Kersigo, J. and Fritzsch, B. (2015). Inner ear hair cells deteriorate in mice engineered to have no or diminished innervation. Frontiers in Aging Neuroscience 7, 33.

Kéver, L., Parmentier, E., Bass, A.H. and Chagnaud, B.P. (2021). Morphological diversity of acoustic and electric communication systems of mochokid catfish. J Comp Neurol 529, 1787–1809.

Kim, E.J., Hori, K., Wyckoff, A., Dickel, L.K., Koundakjian, E.J., Goodrich, L.V. and Johnson, J.E. (2011). Spatiotemporal fate map of neurogenin1 (Neurog1) lineages in the mouse central nervous system. J Comp Neurol 519, 1355–1370.

Kim, W.-Y., Fritzsch, B., Serls, A., Bakel, L.A., Huang, E.J., Reichardt, L.F., Barth, D.S. and Lee, J.E. (2001). NeuroD-null mice are deaf due to a severe loss of the inner ear sensory neurons during development. Development 128, 417–426.

King, A.J. (2020). Feedback systems: Descending pathways and adaptive coding in the auditory system. *In:* The Senses: A Comprehensive Reference (Second Edition), B. Fritzsch, ed. (Oxford: Elsevier), pp. 732–748.

King, A.J. and Walker, K.M. (2020). Listening in complex acoustic scenes. Current Opinion in Physiology 18, 63–72.

Kopecky, B. and Fritzsch, B. (2011). Regeneration of hair cells: Making sense of all the noise. Pharmaceuticals 4, 848–879.

Kopp-Scheinpflug, C. and Forsythe, I.D. (2018). Integration of synaptic and intrinsic conductances shapes microcircuits in the superior olivary complex. *In:* The Mammalian Auditory Pathways (Springer), pp. 101–126.

Kopp-Scheinpflug, C. and Linden, J.F. (2020). Coding of temporal information. *In:* The Senses: A Comprehensive Reference (Second Edition), B. Fritzsch, ed. (Oxford: Elsevier), pp. 691–712.

Kral, A. and Pallas, S.L. (2011). Development of the auditory cortex. *In:* The Auditory Cortex (Springer), pp. 443–463.

Kral, A., Dorman, M.F. and Wilson, B.S. (2019). Neuronal development of hearing and language: Cochlear implants and critical periods. Annu Rev Neurosci 42, e65.

Kral, A., Yusuf, P.A. and Land, R. (2017). Higher-order auditory areas in congenital deafness: Top-down interactions and corticocortical decoupling. Hearing Research 343, 50–63.

Krasewicz, J. and Yu, W.M. (2023). Eph and ephrin signaling in the development of the central auditory system. Developmental Dynamics 252, 10–26.

Kucker, S.C., McMurray, B. and Samuelson, L.K. (2015). Slowing down fast mapping: Redefining the dynamics of word learning. Child Development Perspectives 9, 74–78.

Kumar, S., Gander, P.E., Berger, J.I., Billig, A.J., Nourski, K.V., Oya, H., Kawasaki, H., Howard III, M.A. and Griffiths, T.D. (2021). Oscillatory correlates of auditory working memory examined with human electrocorticography. Neuropsychologia *150*, 107691.

Lang, H. (2016). Loss, degeneration, and preservation of the spiral ganglion neurons and their processes. The Primary Auditory Neurons of the Mammalian Cochlea, 229–262.

Lewis, E.R., Leverenz, E.L. and Bialek, W.S. (1985). Vertebrate Inner Ear. CRC-Press.

Lipovsek, M. and Wingate, R.J. (2018). Conserved and divergent development of brainstem vestibular and auditory nuclei. Elife *7*, e40232.

Lissmann, H.W. (1958). On the function and evolution of electric organs in fish. The Journal of Experimental Biology *35*, 156–191.

Liu, M., Pleasure, S.J., Collins, A.E., Noebels, J.L., Naya, F.J., Tsai, M.-J. and Lowenstein, D.H. (2000). Loss of BETA2/NeuroD leads to malformation of the dentate gyrus and epilepsy. Proceedings of the National Academy of Sciences *97*, 865–870.

Liu, X., Zhang, O., Chen, A., Hu, K., Ehret, G. and Yan, J. (2019). Corticofugal augmentation of the auditory brainstem response with respect to cortical preference. Frontiers in Systems Neuroscience *13*, 39.

Livingston, G., Huntley, J., Sommerlad, A., Ames, D., Ballard, C., Banerjee, S., Brayne, C., Burns, A., Cohen-Mansfield, J. and Cooper, C. (2020). Dementia prevention, intervention, and care: 2020 report of the Lancet Commission. The Lancet *396*, 413–446.

Lombard, R.E. and Hetherington, T.E. (1993). Structural basis of hearing and sound transmission. *In:* The Skull, B.K. Hall and J. Hanken, eds. (Chicago: Chicago University Press), pp. 241–302.

Lowenstein, E.D., Cui, K. and Hernandez-Miranda, L.R. (2022). Regulation of early cerebellar development. FEBS Journal. doi: 10.1111/febs.16426. Epub ahead of print. PMID: 35262281.

Lu, H.-W., Smith, P.H. and Joris, P.X. (2022). Mammalian octopus cells are direction selective to frequency sweeps by excitatory synaptic sequence detection. Proceedings of the National Academy of Sciences *119*, e2203748119.

Luo, Z.-X. and Manley, G.A. (2020). Origins and early evolution of mammalian ears and hearing function. *In:* The Senses: A Comprehensive Reference (Second Edition), B. Fritzsch, ed. (Oxford: Elsevier), pp. 207–252.

Lysakowski, A. (2021). Anatomy and microstructural organization of vestibular hair cells. *In:* The Senses, B. Fritzsch, ed. (Elsevier), pp. 173–184.

Macova, I., Pysanenko, K., Chumak, T., Dvorakova, M., Bohuslavova, R., Syka, J., Fritzsch, B. and Pavlinkova, G. (2019). Neurod1 is essential for the primary tonotopic organization and related auditory information processing in the midbrain. Journal of Neuroscience *39*, 984–1004.

Maklad, A. and Fritzsch, B. (2003). Development of vestibular afferent projections into the hindbrain and their central targets. Brain Research Bulletin *60*, 497–510.

Malmierca, M.S. (2015). Auditory system. *In:* The Rat Nervous System (Elsevier), pp. 865–946.

Malone, B.J., Hasenstaub, A.R. and Schreiner, C.E. (2020). Primary auditory cortex II. Some functional considerations. *In:* The Senses: A Comprehensive Reference (Second Edition), B. Fritzsch, ed. (Oxford: Elsevier), pp. 657–680.

Manley, G.A. (2018). Travelling waves and tonotopicity in the inner ear: A historical and comparative perspective. Journal of Comparative Physiology A 204, 773–781.

Manley, G.A. and Sienknecht, U.J. (2013). The evolution and development of middle ears in land vertebrates. *In:* The Middle Ear (Springer), pp. 7–30.

Manteuffel, G. and Naujoks-Manteuffel, C. (1990). Anatomical connections and electrophysiological properties of toral and dorsal tegmental neurons in the terrestrial urodele *Salamandra salamandra*. Journal fur Hirnforschung 31, 65–76.

Manuel, M., Tan, K.B., Kozic, Z., Molinek, M., Marcos, T.S., Razak, M.F.A., Dobolyi, D., Dobie, R., Henderson, B.E. and Henderson, N.C. (2022). Pax6 limits the competence of developing cerebral cortical cells to respond to inductive intercellular signals. PLoS Biology 20, e3001563.

Maricich, S.M., Xia, A., Mathes, E.L., Wang, V.Y., Oghalai, J.S., Fritzsch, B. and Zoghbi, H.Y. (2009). Atoh1-lineal neurons are required for hearing and for the survival of neurons in the spiral ganglion and brainstem accessory auditory nuclei. Journal of Neuroscience 29, 11123–11133.

Mastick, G.S., Fan, C.M., Tessier-Lavigne, M., Serbedzija, G.N., McMahon, A.P. and Easter, S.S., Jr. (1996). Early deletion of neuromeres in Wnt-1-/- mutant mice: Evaluation by morphological and molecular markers. J Comp Neurol 374, 246–258.

Matei, V., Pauley, S., Kaing, S., Rowitch, D., Beisel, K., Morris, K., Feng, F., Jones, K., Lee, J. and Fritzsch, B. (2005). Smaller inner ear sensory epithelia in Neurog1 null mice are related to earlier hair cell cycle exit. Developmental Dynamics 234, 633–650.

Matesz, C. (1979). Central projection of the VIIIth cranial nerve in the frog. Neuroscience 4, 2061–2071.

McGurk, H. and MacDonald, J. (1976). Hearing lips and seeing voices. Nature 264, 746–748.

McMurray, B., Horst, J.S. and Samuelson, L.K. (2012). Word learning emerges from the interaction of online referent selection and slow associative learning. Psychological Review 119, 831.

Mehr, S.A., Krasnow, M.M., Bryant, G.A. and Hagen, E.H. (2020). Origins of music in credible signaling. Behav Brain Sci 44, e60.

Michalski, N. and Petit, C. (2019). Genes involved in the development and physiology of both the peripheral and central auditory systems. Annual Review of Neuroscience 42, 67–86.

Michalski, N. and Petit, C. (2022). Central auditory deficits associated with genetic forms of peripheral deafness. Human Genetics 141, 335–345.

Miescher, F. (1869). Letter i; to wilhelm his; tübingen, february 26th, 1869. Die histochemischen und physiologischen arbeiten von Friedrich Miescher-aus dem wissenschaftlichen Briefwechsel von F Miescher 1, 33–38.

Milinkeviciute, G. and Cramer, K. (2020). Development of the ascending auditory pathway. *In:* The Senses: A Comprehensive Reference (Second Edition), Fritzsch, B., ed. Vol. 2, pp. 337–353.

Millot, J. and Anthony, J. (1965). Anatomie de *Latimeria chalumnae*. 2. Système nerveux et organes des sens (Ed. du Centre National de la Recherche Scientifique).

Miyata, T., Maeda, T. and Lee, J.E. (1999). NeuroD is required for differentiation of the granule cells in the cerebellum and hippocampus. Genes & Development *13*, 1647–1652.

Molnár, Z., Luhmann, H.J. and Kanold, P.O. (2020). Transient cortical circuits match spontaneous and sensory-driven activity during development. Science *370*, eabb2153.

Moser, T. (2020). Presynaptic physiology of cochlear inner hair cells. *In:* The Senses: A Comprehensive Reference (Second Edition), B. Fritzsch, ed. (Oxford: Elsevier), pp. 441–467.

Mukherjee, D., Meng, X., Kao, J.P.Y. and Kanold, P.O. (2021). Impaired hearing and altered subplate circuits during the first and second postnatal weeks of otoferlin-deficient mice. Cerebral Cortex *32*, 2816–2830.

Müller, N.I., Paulußen, I., Hofmann, L.N., Fisch, J.O., Singh, A. and Friauf, E. (2022). Development of synaptic fidelity and action potential robustness at an inhibitory sound localization circuit: Effects of otoferlin-related deafness. The Journal of Physiology *600*, 2461–2497.

Müller, N.I., Sonntag, M., Maraslioglu, A., Hirtz, J.J. and Friauf, E. (2019). Topographic map refinement and synaptic strengthening of a sound localization circuit require spontaneous peripheral activity. The Journal of Physiology *597*, 5469–5493.

Mulry, E. and Parham, K. (2020). Inner ear proteins as potential biomarkers. Otology & Neurotology *41*, 145–152.

Muniak, M.A., Connelly, C.J., Suthakar, K., Milinkeviciute, G., Ayeni, F.E. and Ryugo, D.K. (2016). Central projections of spiral ganglion neurons. *In:* The Primary Auditory Neurons of the Mammalian Cochlea (Springer), pp. 157–190.

Murray, M.M., Molholm, S., Michel, C.M., Heslenfeld, D.J., Ritter, W., Javitt, D.C., Schroeder, C.E. and Foxe, J.J. (2005). Grabbing your ear: Rapid auditory–somatosensory multisensory interactions in low-level sensory cortices are not constrained by stimulus alignment. Cerebral Cortex *15*, 963–974.

Nakamura, H. (2020). Midbrain patterning: Polarity formation of the tectum, midbrain regionalization, and isthmus organizer. *In:* Patterning and Cell Type Specification in the Developing CNS and PNS (Elsevier), pp. 87–106.

Nakamura, T., Schneider, I. and Shubin, N.H. (2021). Evolution: The Deep Genetic Roots of Tetrapod-specific Traits. Current Biology *31*, R467-R469.

Nakano, Y., Wiechert, S., Fritzsch, B. and Bánfi, B. (2020). Inhibition of a transcriptional repressor rescues hearing in a splicing factor-deficient mouse. Life Science Alliance *3(12)*, e202000841.

Narins, P.M., Feng, A.S. and Fay, R.R. (2006). Hearing and Sound Communication in Amphibians, Vol 28. Springer Science & Business Media.

Newman, E.A., Kim, D.W., Wan, J., Wang, J., Qian, J. and Blackshaw, S. (2018). Foxd1 is required for terminal differentiation of anterior hypothalamic neuronal subtypes. Developmental Biology *439*, 102–111.

Newman, S.A. and Bhat, R. (2009). Dynamical patterning modules: A "pattern language" for development and evolution of multicellular form. International Journal of Developmental Biology *53*, 693–705.

Ngodup, T., Romero, G.E. and Trussell, L.O. (2020). Identification of an inhibitory neuron subtype, the L-stellate cell of the cochlear nucleus. Elife *9*, e54350.

Nichols, D.H., Pauley, S., Jahan, I., Beisel, K.W., Millen, K.J. and Fritzsch, B. (2008). Lmx1a is required for segregation of sensory epithelia and normal ear histogenesis and morphogenesis. Cell and Tissue Research 334, 339–358.

Nieuwenhuys, R. and Puelles, L. (2015). Towards a New Neuromorphology. Springer.

Niklas, K.J. (2014). The evolutionary-developmental origins of multicellularity. American Journal of Botany 101, 6–25.

Ninkovic, J. and Götz, M. (2015). How to make neurons – Thoughts on the molecular logic of neurogenesis in the central nervous system. Cell Tissue Res 359, 5–16.

Northcutt, R.G., Brändle, K. and Fritzsch, B. (1995). Electroreceptors and mechanosensory lateral line organs arise from single placodes in axolotls. Developmental Biology 168, 358–373.

O'Neill, P., Mak, S.-S., Fritzsch, B., Ladher, R.K. and Baker, C.V. (2012). The amniote paratympanic organ develops from a previously undiscovered sensory placode. Nature Communications 3, 1041.

O'Sullivan, A.E., Crosse, M.J., Di Liberto, G.M., de Cheveigné, A. and Lalor, E.C. (2021). Neurophysiological indices of audiovisual speech processing reveal a hierarchy of multisensory integration effects. Journal of Neuroscience 41, 4991–5003.

Oertel, D. and Cao, X.-J. (2020). The ventral cochlear nucleus. *In:* The Senses: A Comprehensive Reference (Second Edition), B. Fritzsch, ed. (Oxford: Elsevier), pp. 517–532.

Ohno, S. (1970). Duplication for the sake of producing more of the same. *In:* Evolution by Gene Duplication (Springer), pp. 59–65.

Ohno, S. (2013). Evolution by Gene Duplication. Springer Science & Business Media.

Overstreet-Wadiche, L.S., Bensen, A.L. and Westbrook, G.L. (2006). Delayed development of adult-generated granule cells in dentate gyrus. Journal of Neuroscience 26, 2326–2334.

Owen, R. (1848). On the archetype and homologies of the vertebrate skeleton (author). London, John van Voorst, pp. 1–203.

Pan, N., Kopecky, B., Jahan, I. and Fritzsch, B. (2012). Understanding the evolution and development of neurosensory transcription factors of the ear to enhance therapeutic translation. Cell and Tissue Research 349, 415–432.

Patton, M.H., Blundon, J.A. and Zakharenko, S.S. (2019). Rejuvenation of plasticity in the brain: Opening the critical period. Current Opinion in Neurobiology 54, 83–89.

Pauley, S., Lai, E. and Fritzsch, B. (2006). Foxg1 is required for morphogenesis and histogenesis of the mammalian inner ear. Developmental dynamics: An official publication of the American Association of Anatomists 235, 2470–2482.

Pawlicki, M., Collins, H.A., Denning, R.G. and Anderson, H.L. (2009). Two-photon absorption and the design of two-photon dyes. Angewandte Chemie International Edition 48, 3244–3266.

Pecka, M. and Encke, J. (2020). Coding of Spatial Information. *In:* The Senses: A Comprehensive Reference (Second Edition), B. Fritzsch, ed. (Oxford: Elsevier), pp. 713–731.

Peter, I.S. and Davidson, E.H. (2015). Genomic Control Process: Development and Evolution (Academic Press).

Petitpré, C., Faure, L., Uhl, P., Fontanet, P., Filova, I., Pavlinkova, G., Adameyko, I., Hadjab, S. and Lallemend, F. (2022). Single-cell RNA-sequencing analysis of the developing mouse inner ear identifies molecular logic of auditory neuron diversification. Nature Communications 13, 1–15.

Pheasant, R.J., Fisher, M.N., Watts, G.R., Whitaker, D.J. and Horoshenkov, K.V. (2010). The importance of auditory-visual interaction in the construction of 'tranquil space'. Journal of Environmental Psychology 30, 501–509.

Phillips-Silver, J. and Trainor, L.J. (2008). Vestibular influence on auditory metrical interpretation. Brain and Cognition 67, 94–102.

Phillips-Silver, J., VanMeter, J.W. and Rauschecker, J.P. (2020). Auditory-Vestibulomotor Temporal Processing and Crossmodal Plasticity for Musical Rhythm in the Early Blind. bioRxiv.

Phillips, J.O., Ling, L., Nowack, A., Rebollar, B. and Rubinstein, J.T. (2020). Interactions between auditory and vestibular modalities during stimulation with a combined vestibular and cochlear prosthesis. Audiology and Neurotology 25, 96–108.

Pierce, M.L., Weston, M.D., Fritzsch, B., Gabel, H.W., Ruvkun, G. and Soukup, G.A. (2008). MicroRNA-183 family conservation and ciliated neurosensory organ expression. Evolution & Development 10, 106-113.

Plass, J., Brang, D., Suzuki, S. and Grabowecky, M. (2020). Vision perceptually restores auditory spectral dynamics in speech. Proceedings of the National Academy of Sciences 117, 16920–16927.

Platt, C., Jørgensen, J.M. and Popper, A.N. (2004). The inner ear of the lungfish *Protopterus*. Journal of Comparative Neurology 471, 277–288.

Polley, D.B., Read, H.L., Storace, D.A. and Merzenich, M.M. (2007). Multiparametric auditory receptive field organization across five cortical fields in the albino rat. Journal of Neurophysiology 97, 3621-3638.

Polley, D.B. and Schiller, D. (2022). The promise of low-tech intervention in a high-tech era: Remodeling pathological brain circuits using behavioral reverse engineering. Neuroscience & Biobehavioral Reviews, 104652.

Polley, D.B. and Takesian, A.E. (2022). Thalamocortical circuits for auditory processing, plasticity, and perception. *In:* The Thalamus, Halassa, M.M. ed. (Cambridge: Cambridge University Press), pp. 237–268.

Popper, A.N., Hawkins, A.D. and Sisneros, J.A. (2021). Fish hearing "specialization" – A re-evaluation. Hearing Research, 108393.

Puelles, L. (2019). Survey of midbrain, diencephalon, and hypothalamus neuroanatomic terms whose prosomeric definition conflicts with columnar tradition. Frontiers in Neuroanatomy 13, 20.

Puelles, L., Martinez-de-la-Torre, M., Ferran, J.-L. and Watson, C. (2012). Diencephalon. *In:* The Mouse Nervous System (Elsevier), pp. 313–336.

Puelles, L., Martínez, S., Martínez-De-La-Torre, M. and Rubenstein, J.L. (2015). Gene maps and related histogenetic domains in the forebrain and midbrain. *In:* The Rat Nervous System (Elsevier), pp. 3–24.

Rauschecker, J.P. (2020). The auditory cortex of primates including man with reference to speech. *In:* The Senses: A Comprehensive Reference (Second Edition), B. Fritzsch, ed. (Oxford: Elsevier), pp. 791–811.

Rauschecker, J.P. and Scott, S.K. (2009). Maps and streams in the auditory cortex: Nonhuman primates illuminate human speech processing. Nat Neurosci 12, 718–724.

Read, H.L. and Reyes, A.D. (2018). Sensing sound through Thalamocortical afferent architecture and cortical microcircuits. *In:* The Mammalian Auditory Pathways (Springer), pp. 169–198.

Rees, A. (2020). The inferior colliculus. *In:* The Senses: A Comprehensive Reference (Second Edition), B. Fritzsch, ed. (Oxford: Elsevier), pp. 566–600.

Reichert, K. (1837). Über die Visceralbogen der Wirbelthiere im Allgemeinen und deren Metamorphosen bei den Vögeln und Säugethieren. Arch Anat Phys Wiss Med *1837*, 120–220.

Retzius, G. (1881). Das Gehörorgan der Fische und Amphibien, Vol 1 (Gedruckt in der Centraldruckerei in Commission bei Samson & Wallin).

Retzius, G. (1884). Das Gehororgan der Saugethiere und des Menschen. Das Gehororgan der Wirbelthiere, Morphologisch-Histologische Studien. II. Das Gehororgan der Reptilien, der Vogel und der Saugethiere, 201–368.

Rolls, E.T., Rauschecker, J.P., Deco, G., Huang, C.-C. and Feng, J. (2022). Auditory cortical connectivity in humans. Cerebral Cortex bhac496. doi: 10.1093/cercor/bhac496. Epub ahead of print. PMID: 36573464.

Rosowski, J.J. (2013). Comparative middle ear structure and function in vertebrates. The Middle Ear, 31–65.

Roux, I., Safieddine, S., Nouvian, R., Grati, M.H., Simmler, M.-C., Bahloul, A., Perfettini, I., Le Gall, M., Rostaing, P. and Hamard, G. (2006). Otoferlin, defective in a human deafness form, is essential for exocytosis at the auditory ribbon synapse. Cell *127*, 277–289.

Rubio, M.E. (2018a). Microcircuits of the ventral cochlear nucleus. *In:* The Mammalian Auditory Pathways (Springer), pp. 41–71.

Rubio, M.E. (2018b). Molecular and Structural Changes in the Cochlear Nucleus in Response to Hearing Loss. The Oxford Handbook of the Auditory Brainstem.

Rubio, M.E. (2020). Auditory brainstem development and plasticity. Current Opinion in Physiology *18*, 7–10.

Rubio, M.E. and Juiz, J.M. (2004). Differential distribution of synaptic endings containing glutamate, glycine, and GABA in the rat dorsal cochlear nucleus. J Comp Neurol *477*, 253–272.

Ryan, M.J. and Guerra, M.A. (2014). The mechanism of sound production in túngara frogs and its role in sexual selection and speciation. Current Opinion in Neurobiology *28*, 54–59.

Ryugo, D.K. and Parks, T.N. (2003). Primary innervation of the avian and mammalian cochlear nucleus. Brain Research Bulletin *60*, 435–456.

Samantsidis, G.-R., Panteleri, R., Denecke, S., Kounadi, S., Christou, I., Nauen, R., Douris, V. and Vontas, J. (2020). 'What I cannot create, I do not understand': Functionally validated synergism of metabolic and target site insecticide resistance. Proceedings of the Royal Society B *287*, 20200838.

Sanders, T.R. and Kelley, M.W. (2022). Specification of neuronal subtypes in the spiral ganglion begins prior to birth in the mouse. Proc Natl Acad Sci USA *119*, e2203935119.

Savage, P.E., Loui, P., Tarr, B., Schachner, A., Glowacki, L., Mithen, S. and Fitch, W.T. (2021). Music as a coevolved system for social bonding. Behavioral and Brain Sciences *44*, e59.

Schinzel, F., Seyfer, H., Ebbers, L. and Nothwang, H.G. (2021). The Lbx1 lineage differentially contributes to inhibitory cell types of the dorsal cochlear nucleus, a cerebellum-like structure, and the cerebellum. Journal of Comparative Neurology *529*, 3032–3045.

Schmidt, H. and Fritzsch, B. (2019). Npr2 null mutants show initial overshooting followed by reduction of spiral ganglion axon projections combined with near-normal cochleotopic projection. Cell and Tissue Research *378*, 15–32.

Schneider, J.N. and Mercado III, E. (2019). Characterizing the rhythm and tempo of sound production by singing whales. Bioacoustics *28*, 239–256.

Schultz, J.A., Zeller, U. and Luo, Z.X. (2017). Inner ear labyrinth anatomy of monotremes and implications for mammalian inner ear evolution. Journal of Morphology *278*, 236–263.

Schulz-Mirbach, T., Ladich, F., Plath, M. and Heß, M. (2019). Enigmatic ear stones: What we know about the functional role and evolution of fish otoliths. Biological Reviews *94*, 457–482.

Shafin, K., Pesout, T., Lorig-Roach, R., Haukness, M., Olsen, H.E., Bosworth, C., Armstrong, J., Tigyi, K., Maurer, N. and Koren, S. (2020). Nanopore sequencing and the Shasta toolkit enable efficient de novo assembly of eleven human genomes. Nature Biotechnology *38*, 1044–1053.

Shupak, A. (2006). Recurrent diving-related inner ear barotrauma. Otology & Neurotology *27*, 1193–1196.

Simmons, A.M. (2019). Tadpole bioacoustics: Sound processing across metamorphosis. Behavioral Neuroscience *133*, 586.

Simmons, D., Duncan, J., de Caprona, D.C. and Fritzsch, B. (2011). Development of the inner ear efferent system. *In:* Auditory and Vestibular Efferents (Springer), pp. 187–216.

Smotherman, M. and Narins, P. (2004). Evolution of the amphibian ear. *In:* Evolution of the Vertebrate Auditory System, G.A. Manley, A.N. Popper and R.R. Fay, eds. (Berlin: Springer), pp. 164–199.

Spence, C. (2011). Crossmodal correspondences: A tutorial review. Attention, Perception, & Psychophysics *73*, 971–995.

Spence, C. and Soto-Faraco, S. (2010). Auditory perception: Interactions with vision. The Oxford Handbook of Auditory Science: Hearing *3*, 271–296.

Spence, C., Wang, Q.J., Reinoso-Carvalho, F. and Keller, S. (2021). Commercializing sonic seasoning in multisensory offline experiential events and online tasting experiences. Front Psychol *12*, 740354.

Steinschneider, M., Nourski, K.V. and Fishman, Y.I. (2013). Representation of speech in human auditory cortex: Is it special? Hearing Research *305*, 57–73.

Stopschinski, B.E., Del Tredici, K., Estill-Terpack, S.-J., Ghebremdehin, E., Yu, F.F., Braak, H. and Diamond, M.I. (2021). Anatomic survey of seeding in Alzheimer's disease brains reveals unexpected patterns. Acta Neuropathologica Communications *9*, 1–19.

Straka, H., Lambert, F.M. and Simmers, J. (2022). Role of locomotor efference copy in vertebrate gaze stabilization. Paper presented at Frontiers in Neural Circuits.

Sun, Y., Wang, L., Zhu, T., Wu, B., Wang, G., Luo, Z., Li, C., Wei, W. and Liu, Z. (2022). Single-cell transcriptomic landscapes of the otic neuronal lineage at multiple early embryonic ages. Cell Reports *38*, 110542.

Suryanarayana, S.M., Pérez-Fernández, J., Robertson, B. and Grillner, S. (2020). The evolutionary origin of visual and somatosensory representation in the vertebrate pallium. Nature Ecology & Evolution *4*, 639–651.

Szeto, I.Y., Chu, D.K., Chen, P., Chu, K.C., Au, T.Y., Leung, K.K., Huang, Y.-H., Wynn, S.L., Mak, A.C. and Chan, Y.-S. (2022). SOX9 and SOX10 control fluid homeostasis in the inner ear for hearing through independent and cooperative mechanisms. Proceedings of the National Academy of Sciences *119*, e2122121119.

Szostakiwskyj, M. and Anderson, J.S. (2022). Salamander braincase morphology as revealed by micro-computed tomography. Journal of Morphology *283*, 462–501.

Tan, R., Wang, Y., Kleinstein, S.E., Liu, Y., Zhu, X., Guo, H., Jiang, Q., Allen, A.S. and Zhu, M. (2014). An evaluation of copy number variation detection tools from whole-exome sequencing data. Human Mutation *35*, 899–907.

Taverna, E., Götz, M. and Huttner, W.B. (2014). The cell biology of neurogenesis: Toward an understanding of the development and evolution of the neocortex. Annual Review of Cell and Developmental Biology *30*, 465–502.

Tichko, P., Kim, J.C. and Large, E.W. (2021). Bouncing the network: A dynamical systems model of auditory–vestibular interactions underlying infants' perception of musical rhythm. Developmental Science, e13103.

Tran, H.N., Nguyen, Q.H., Jeong, J.E., Loi, D.L., Nam, Y.H., Kang, T.H., Yoon, J., Baek, K. and Jeong, Y. (2023). The embryonic patterning gene Dbx1 governs the survival of the auditory midbrain via Tcf7l2-Ap2δ transcriptional cascade. Cell Death & Differentiation, 1-12.

Trudeau-Fisette, P., Ito, T. and Ménard, L. (2019). Auditory and somatosensory interaction in speech perception in children and adults. Frontiers in Human Neuroscience *13*, 344.

Trussell, L.O. and Oertel, D. (2018). Microcircuits of the dorsal cochlear nucleus. *In*: The Mammalian Auditory Pathways (Springer), pp. 73–99.

Tucker, A.S. (2017). Major evolutionary transitions and innovations: The tympanic middle ear. Philosophical Transactions of the Royal Society B: Biological Sciences *372*, 20150483.

Uchida, Y., Nishita, Y., Otsuka, R., Sugiura, S., Sone, M., Yamasoba, T., Kato, T., Iwata, K. and Nakamura, A. (2022). Aging brain and hearing: A mini-review. Frontiers in Aging Neuroscience *13*, 991.

Urbánek, P., Fetka, I., Meisler, M.H. and Busslinger, M. (1997). Cooperation of Pax2 and Pax5 in midbrain and cerebellum development. Proceedings of the National Academy of Sciences *94*, 5703–5708.

van Bergeijk, W.A. (1967). The evolution of vertebrate hearing. Contributions to Sensory Physiology *2*, 1–49.

van Dyck, C.H., Swanson, C.J., Aisen, P., Bateman, R.J., Chen, C., Gee, M., Kanekiyo, M., Li, D., Reyderman, L. and Cohen, S. (2022). Lecanemab in early Alzheimer's disease. New England Journal of Medicine *388(1)*, 9–21.

Vignieri, S. (2023 Apr 28). Zoonomia. Science *380(6643)*, 356-357. doi: 10.1126/science.adi1599. Epub 2023 Apr 27.

von Baer, K.E. (1828). Über entwickelungsgeschichte der thiere: Beobachtung und reflexion, Vol 1 (Bei den gebrüdern Bornträger).

von Bartheld, C.S. (1992). Oculomotor and sensory mesencephalic trigeminal neurons in lungfishes: Phylogenetic implications. Brain, Behavior and Evolution *39*, 247–263.

Wagner, A. (2011). The origins of evolutionary innovations: A theory of transformative change in living systems (Oxford: OUP).

Wagner, A. (2015). Arrival of the Fittest: How Nature Innovates (Current).
Wahnschaffe, U., Bartsch, U. and Fritzsch, B. (1987). Metamorphic changes within the lateral-line system of Anura. Anatomy and Embryology 175, 431–442.
Wake, D.B. and Hanken, J. (1996). Direct development in the lungless salamanders: What are the consequences for developmental biology, evolution and phylogenesis? Int J Dev Biol 40, 859–869.
Wake, M.H. and Donnelly, M.A. (2010). A new lungless caecilian (Amphibia: Gymnophiona) from Guyana. Proceedings of the Royal Society B: Biological Sciences 277, 915–922.
Walton, P.L., Christensen-Dalsgaard, J. and Carr, C.E. (2017). Evolution of sound source localization circuits in the nonmammalian vertebrate brainstem. Brain, Behavior and Evolution 90, 131–153.
Wang, H.-F., Zhang, W., Rolls, E.T., Li, Y., Wang, L., Ma, Y.-H., Kang, J., Feng, J., Yu, J.-T. and Cheng, W. (2022). Hearing impairment is associated with cognitive decline, brain atrophy and tau pathology. Ebiomedicine 86, 104336.
Wang, K., Wang, J., Zhu, C., Yang, L., Ren, Y., Ruan, J., Fan, G., Hu, J., Xu, W. and Bi, X. (2021a). African lungfish genome sheds light on the vertebrate water-to-land transition. Cell 184, 1362–1376. e1318.
Wang, Q.J., Keller, S. and Spence, C. (2021b). Metacognition and crossmodal correspondences between auditory attributes and saltiness in a large sample study. Multisensory Research 34, 785–805.
Wangemann, P. (2002). K^+ cycling and the endocochlear potential. Hearing Research 165, 1–9.
Watson, C., Shimogori, T. and Puelles, L. (2017). Mouse Fgf8-Cre-LacZ lineage analysis defines the territory of the postnatal mammalian isthmus. Journal of Comparative Neurology 525, 2782–2799.
Wenstrup, J.J., Ghasemahmad, Z., Hazlett, E. and Shanbhag, S.J. (2020). The Amygdala – A hub of the social auditory brain. *In:* The Senses: A Comprehensive Reference (Second Edition), B. Fritzsch, ed. (Oxford: Elsevier), pp. 812–837.
Wichova, H., Alvi, S., Boatright, C., Ledbetter, L., Staecker, H. and Lin, J. (2019). High-resolution computed tomography of the inner ear: Effect of otosclerosis on cochlear aqueduct dimensions. Annals of Otology, Rhinology & Laryngology 128, 749–754.
Winer, J.A. and Schreiner, C.E. (2011). Toward a synthesis of cellular auditory forebrain functional organization. *In:* The Auditory Cortex (Springer), pp. 679–686.
Witschi, E. (1949). The larval ear of the frog and its transformation during metamorphosis. Zeitschrift für Naturforschung B 4, 230–242.
Wollesen, T., Rodríguez Monje, S.V., Todt, C., Degnan, B.M. and Wanninger, A. (2015). Ancestral role of Pax2/5/8 in molluscan brain and multimodal sensory system development. BMC Evolutionary Biology 15, 1–19.
Wu, P.Z., Liberman, L.D., Bennett, K., de Gruttola, V., O'Malley, J.T. and Liberman, M.C. (2019). Primary neural degeneration in the human cochlea: Evidence for hidden hearing loss in the aging ear. Neuroscience 407, 8–20.
Xu, J., Li, J., Zhang, T., Jiang, H., Ramakrishnan, A., Fritzsch, B., Shen, L. and Xu, P.X. (2021). Chromatin remodelers and lineage-specific factors interact to target enhancers to establish proneurosensory fate within otic ectoderm. PNAS 118, 1–12.

Yu, X. and Wang, Y. (2022). Tonotopic differentiation of presynaptic neurotransmitter-releasing machinery in the auditory brainstem during the prehearing period and its selective deficits in Fmr1 knockout mice. Journal of Comparative Neurology 530.18, 3248–3269.

Zampini, M. and Spence, C. (2012). Assessing the role of visual and auditory cues in multisensory perception of flavor. *In:* The Neural Bases of Multisensory Processes. CRC Press/Taylor & Francis, Boca Raton (FL). PMID: 22593877.

Zeyl, J.N., den Ouden, O., Köppl, C., Assink, J., Christensen-Dalsgaard, J., Patrick, S.C. and Clusella-Trullas, S. (2020). Infrasonic hearing in birds: A review of audiometry and hypothesized structure–function relationships. Biological Reviews 95, 1036–1054.

Zeyl, J.N., Snelling, E.P., Connan, M., Basille, M., Clay, T.A., Joo, R., Patrick, S.C., Phillips, R.A., Pistorius, P.A. and Ryan, P.G. (2022). Aquatic birds have middle ears adapted to amphibious lifestyles. Scientific Reports 12, 1–12.

Zhang, M., Kwon, S.E., Ben-Johny, M., O'Connor, D.H. and Issa, J.B. (2020). Spectral hallmark of auditory-tactile interactions in the mouse somatosensory cortex. Communications Biology 3, 1–17.

Zhao, J., Xu, W. and Ye, L. (2018). Effects of auditory-visual combinations on perceived restorative potential of urban green space. Applied Acoustics 141, 169–177.

Zhu, Y.-A., Giles, S., Young, G.C., Hu, Y., Bazzi, M., Ahlberg, P.E., Zhu, M. and Lu, J. (2021). Endocast and bony labyrinth of a Devonian "Placoderm" challenges stem gnathostome phylogeny. Current Biology 31, 1112–1118. e1114.

Zine, A. and Fritzsch, B. (2023). Early steps towards hearing: Placodes and sensory development. International Journal of Molecular Sciences 24(8), 699.

Zuo, J., Treadaway, J., Buckner, T.W. and Fritzsch, B. (1999). Visualization of α9 acetylcholine receptor expression in hair cells of transgenic mice containing a modified bacterial artificial chromosome. Proceedings of the National Academy of Sciences 96, 14100–14105.

Index

A

A1, auditory field 1, 159, 165, 170
AAF, associated auditory field, 159, 165
AC, anterior canal (crista), 55, 60, 69, 109, 131
AC, auditory cortex, 155-157, 159-163, 165, 169, 175, 176, 178, 179, 181, 184
Actinopterygians, 93, 97, 101, 172
Agnathans, 102
Air, 101
aLL, Anterior lateral line, 55, 73-76, 82
Alzheimer's disease, 182-184
Amniotes, 93, 95, 97, 99, 108, 112, 128, 185
AMPA, a-amino-3-hydroxy-5-methyl-4-isoxazoleprpionic acid, 157, 163
Amphibian papilla, 69, 94, 101, 112, 172, 185
Amphibians, 73, 80, 92, 93, 96-66, 101, 108, 110, 123, 128, 173, 185
Ampullae of Lorenzini, 87, 90
Ampullary organs (AO), 83
Amygdala, 184
Amyloid, 183-184
And1-3, actinodin1-3, 97
Angular orientation, 52
Anurans, 78, 82, 84, 92
Apex, 131, 132, 173
Area, 27, 35, 46, 51
Artificial intelligence, AI, 3

Ascl1, achaete-scute family bHLH transcription factor 1, 53, 64, 153, 159
Atoh1, atonal bHLH transcription factor 1, 22, 28 36, 38-41, 45, 55, 66, 71, 73, 76, 86, 87, 121, 124, 130- 138
ATP, 9
Attenuation, 181
Auditory development, 95
Auditory nuclei, 115, 121
Auditory sensory organs, 94
Auditory system, 51, 58, 93-95, 113-116, 171
Auditory-somatic interactions, 180
Auditory-taste interactions, 181
Auditory-vestibular interactions, 181
Auditory-visual interactions, 180
AVCN, anteroventral cochlear nucleus, 55, 132, 138-140, 151
AVST, anterior (superior) vestibulospinal tract, 64

B

Bacteria, 6, 8
Base pairs, 8
Base, 131
Basic helix-loop-helix (bHLH) genes, 16, 20
Basicranial muscle, 104
Basilar membrane, 104, 105, 107, 185
Basilar papilla, 69, 93-95, 99, 101, 103,

Index

106, 108, 110, 112, 123, 128, 129, 133, 135, 171-174, 185
Bats, 2, 148, 164
Belt (caudal, surrounding), 178-179
BIC, brachium of the inferior colliculus, 153
Bichir, 82, 84, 89, 90, 92, 94, 96, 99, 108
Bilaterians, 17, 23, 46
Biology, 1, 2, 7
Birds, 2, 175
BMPs, bone morphogenetic proteins, 86, 96, 163
Bony fish, 95, 129
Brain waves, 3
Brain-derived neurotrophic factor (BDNF), 49, 91, 119, 125
Brainstem, 51, 56, 58-60, 177
Broca's area, 179, 186
Brodmann's area, 178
Bushy cells, 139-141, 143, 145

C

CA1-3, cornu ammonis, 182-183
Caecilians (gymnophionans), 75, 78, 82, 83, 89, 90, 92, 93, 98, 101, 103, 110, 112, 116, 129, 135
Cajal-Retzius, 162-165
Cartwheel cells, 142, 143, 145
Cdh23, cadherin related 23, 11, 12, 70, 78, 126
Cell cycle, 16, 20, 21
Cell nuclei, 4
Cellular morphology, 15
Central cochlear nuclei, 116
Central nervous system (CNS), 31, 43, 50, 114
Central projections, 55, 76, 93, 114, 130, 134, 167, 174
ChatGPT, 3
Choanae, 107
Choanoflagellate, 17, 24, 26, 53, 73, 125
Chondrichthyes, cartilaginous fish, 69, 74, 79, 82, 84, 89, 92, 102
Chordates, 49-52, 56, 115
Choroid plexus, 56, 63, 66, 137, 146
Chromosome, 6, 20
Cib2, calcium and integrin binding family member 2, 12-14, 28, 126

CN, cochlear nuclei, 54, 76, 114, 146, 174, 186
Cnidarians, 23
Cochlea, 93, 118, 123
Cochlear aqueduct, 93, 99, 105-108, 110, 172, 173, 185
Cochlear hair cells, 124
Coelacanth, 97, 103, 105
Coelenterate, 31
Comb jellies, 31, 35
Constant frequency (CF), 169
Cortical plasticity, 167-169
Corticocortical axons, projections, connections, 165-166
Corticothalamic connections, 165-166
Coughing, 177
Craniates, 27, 48, 56, 82
CSF, Cerebrospinal fluid, 110, 172, 185
Cycle, 6, 16

D

DC, dorsal crista, 60, 69
DCN, dorsal cochlear nucleus, 55, 134, 148, 174
Dentate gyrus, 182, 183
Deuterostomia, 25, 115
Diffusion tractography, 179
DN, dorsal nucleus, 54, 85
DNA, deoxyribonucleic acid, 4-10, 16, 22
DNLL, dorsal nucleus of the lateral lemniscus, 140, 147, 150-153, 155
DON, dorsal octavo-lateral nucleus, 82
D-stellate cells, 137, 140, 143, 144

E

Ear, 2, 95, 99, 115
ECoG, electrocorticography, 170
Egr2, early growth response 2 (Krox20), 130, 132, 133, 138
Elasmobranch, 95, 99, 103, 171
Electric fish, 81
Electric signals, 33, 82
Electroreception, 58, 81-84, 89-91, 93, 94, 115, 119, 121, 133-135, 171, 174
Electroreceptive sensory neurons, 82
ELL, Electrosensory lateral line, 55, 82, 84, 86

Embryonic origin, 123
Eminentia granularis, 85
Emx2, empty spiracles homeobox 2, 68, 70, 80, 124
Endolymph, 14, 110, 127, 172
Eustachian tube, 98, 99, 101, 107
Evolution (of molecules), 2-10, 12-15, 23, 89, 91, 95, 99, 171, 173, 176, 177
Evolvability, 5
Eya1, eyes absent 1, 37, 39, 53, 62, 68, 114, 117-119, 124
Eyes, 34-37

F

Far field, 95
Fgf3/10, fibroblast growth factor 3/10, 67, 114
Fgf8, fibroblast growth factor 8, 57, 124, 153
Fish, flying fish, 2
Forebrain, 159, 163
Foxd1, forkhead box d1, 159
Foxg1, forkhead box g1, 57, 62, 68, 117, 124, 159, 163
Foxi3, forkhead box i3, 67, 68
Foxp1, forkhead box p1/2, 97, 159
Frequency (response area, FRA), 169
Frogs, 100, 101, 107, 110, 171-176, 185
Fusiform cells (pyramidal cells), 139, 140, 142, 145

G

GABA, 138-143, 147, 151, 152, 157, 164
Gas bladder, 95, 107
Gata3, GATA binding protein 3, 63, 114, 117, 124
GBCs, see Globose basal cells
Gbx2, gastrulation brain homeobox 2, 57, 154
Gdf7, growth differentiation factor 7, 53, 58, 114, 123, 131
Genes, 6, 38-39
Genetic, 6, 16, 19
Genome, 5, 6
Giant cells, 141, 145
Gill arch, 103, 104
Globular bushy cells (GBCs), 137-140, 143, 147

Glutamate, 138-142, 145, 151, 152, 157
Glycine, 138-140, 142-144, 151, 152
Gnathostomes, 59, 66, 71-75, 90, 93, 102, 104, 171, 174, 175
Golgi cells, 142, 144
Granule cells, 141, 144
Gravistatic receptors, 32, 35, 52-53, 59, 61, 68-70, 108, 123
GRN, gene regulatory network, 20, 37, 38, 40, 41
Gymnotids, 82-84, 89

H

Hagfish, 58-60, 69, 73, 78, 84, 102
Hair cells, mechanosensory HC, 2, 12-15, 26, 36, 48, 68, 109, 114, 173, 185
HC, horizontal canal, 62, 69, 131
Hearing system, 95
Hearing, 2, 93, 172, 182, 185-187
Hippocampus, 182, 184, 185
Hoxb1, homeobox b1, 132, 137
Human brain, 3, 178
Human genome, 5
Human societies, 2
Humans, 2-8, 22, 175, 176, 186, 187
Hydrodynamic function, 176
Hyomandibular bone, 100-104, 173
Hz, kHz, (kilo) Hertz, 12, 89, 111, 158, 174

I

IC, inferior colliculus, 116, 151, 153, 156, 165, 180
ICc, ovoid, central nucleus of the IC, 153, 155
ICd, dorso-medial cortex of the IC, 153, 155
ICl, ventro-lateral cortex of the IC, 153, 155
ILD, interaural (sound) level differences, 149, 152, 157, 168, 175, 186
Impedance, 107
IN, Intermediate nuclei, 54, 75
Inheritance, 7
INLL, intermediate nucleus of the lateral lemniscus, 140, 150-152, 156

Index

Inner ear, 15, 43, 58, 99
Inner hair cells (IHC), 114, 120, 124, 127, 173, 186
Interneurons, 31, 67, 84, 162
Intracranial joint, IJ, 103, 104, 112, 173
Ion channels, 30
IPD, interaural phase difference, 148-150, 157, 176, 186
Irx3, Iroquois related homeobox 3, 57, 159
Isl1, ISL1 transcription factor, 118, 120, 130, 167
Isofrequency, 167
ITD, interaural time difference, 147, 149, 152, 157, 175, 186
IVST, inferior vestibulospinal tract, 64

J

Jellyfish, 31-35

K

K^+ (channels), 10
Kinocilium, 24-28, 40, 53, 69, 89

L

Lagena, 61, 68, 70, 94, 105, 110, 112
Lampreys, 58-60, 69, 73, 78, 80, 82, 84-86, 89, 102, 104, 171
Lancelet, see Amphioxus, 56
Language, 2, 3, 164, 176, 177, 179
Larynx, 95, 176
Lateral lemniscus (LL), 121, 135, 139, 150-153, 180
Lateral line (LL, LLN), 57, 72, 83-86, 90, 93-95, 114-116, 119, 133-135
Lateral olivocochlear bundle (LOC), 120
Latimeria, 69, 82, 84, 89, 90, 95-103, 106-109, 123, 124, 129, 171-173, 186
Layer, 23, 85, 141, 142, 162-164
Learning, 3, 177, 187
LGN, lateral geniculate nucleus, 160, 163
Lmx1a/b, LIM homeobox transcription factor 1 alpha, beta, 54, 57, 62, 66, 68,76, 86, 87, 114, 117, 123-124, 131, 136, 153, 154

lncRNA, long non-coding RNA, 5, 23
LNTB, lateral nuclear of the trapezoid body, 139, 147, 149, 151
LSO, lateral superior olive, 139, 146, 147, 149, 155-157, 168
L-stellate cells, 140, 145
LUCA, last universal common ancestor, 8, 10
Lungfish, 69, 82-84, 89, 90, 92-101, 103-104, 108, 171
Lungs, 93, 95-99, 105, 172
LVST (LV), Lateral vestibulospinal tract, 64-66

M

Mammals, mammalian, 95, 100-102, 107, 110, 123, 172-175, 185
Mechanical stimuli, 2
Mechanotransduction (channels), 9, 12-16, 19, 25-29, 45
Medial geniculate body (MGB), 141, 153, 159, 180, 186
Meis2, meis homeobox 2, 153, 159
Membrane, 12
MET, mechano-electric transducer, 13, 70, 78
Metazoans, 6, 16, 17, 19, 23, 29, 33, 35
Mice, 47, 164
Microvilli, 25, 53,
Midbrain-hindbrain boundary (MHB), 57, 115, 159
miR, microRNA, 5, 23, 26, 68, 88, 115, 124
MNTB, medial nuclear trapezoid body, 139, 146-151, 168
MOC, medial olivocochlear bundle (MOC), 120
Molecules, 2, 5, 8-10, 14, 19
MON, medial octavolateralis nuclei, 75
Monotremes, 110, 124, 129, 173
Mormyrids, 82-87, 89
MSO, medial superior olive, 139, 143, 146-150, 155-157
Multicellular organisms, 16
Musicality, 177
Mutation, 5
MVST, medial vestibulospinal tract, 64

N

Neglected papilla, papilla neglecta, 61, 94, 101, 108, 109
Neurod1, neuronal differentiation 1, 19, 20, 36, 44-48, 53, 63, 68, 72, 117, 124, 130, 133, 153, 154, 159, 167, 182
Neurog1, neurogenin 1, 20, 25, 27, 39, 41, 44-48, 53-56, 59, 63, 65, 71-73, 114, 117, 125, 131, 153, 154
Neurog2, neurogenin 2, 54-55, 65, 153, 154, 159, 163
Neuromasts, 71, 73, 74, 83, 88
Neuronal plasticity, 3
Neurosensory formation, evolution, 32, 73
NICD, notch intracellular domain, 21
Nkx2.1, 96, 97
NMDA, n-methyl-d-aspartate, 157, 163
n-Myc, mycn proto-oncogeme, bHLH transcription factor, 62, 68, 124
Notochord, 104-106, 112
Notopterids (African knifefish), 86
Ntf3 (NT3), neurotrophin 3, 119, 125

O

Octopus cells, 140, 143, 148
Olig3, oligodendrocyte transcription factor 3, 55, 64, 65, 135
Opb, open brain, 153
Operculum, 103
Organ of Corti, 25, 28, 46, 112, 172-174
Organisms, 6
Osteichthyes, 98, 103
Otof, otoferlin, 167, 168
Otx2, orthodenticle homeobox 2, 57, 114, 153
Ouabain, 167, 168
Outer hair cells (OHC), 114, 120, 124, 126, 186
Oval window, 94, 99, 103, 110, 112, 172

P

Pax2, Pax5, Pax6, Pax8, paired box 2/5/6/8, 36-39, 57, 64, 68, 114, 117, 125, 153, 154
PC, posterior crista, 61, 69, 94, 107, 131

Pcdh15, protocadherin related 11-12, 15, 78, 126
PCP, see Planar cell polarity, 70
Perilymph, 172, 185
Perilymphatic space, (PS) 94, 99, 101, 108, 112
Peripheral hair cells, 116, 119
Peripheral nervous system (PNS), 95
Phox2b, paired like homeobox 2b, 65, 66, 136
Placode and neural crest formation, 56, 67
Placozoans, 31
pLL, posterior lateral line, 55, 73-76
Pressure, 99, 103, 106, 123, 173
Proteins, 8-10, 12-16
Protostomia, 26
Prph1, peripherin, 118
PSTH, peri stimulus time histogram, 150
Pterosaurs, 2
Ptf1a, pancreas associated transcription factor 1a, 55, 64, 65, 77, 86, 114, 121, 132, 135-138, 150
PVCN, posteroventral cochlear nuclei, 131, 136, 137, 139, 148, 151, 156
Pyramidal neurons, 162-165, 168

R

r0-11, rhombomeres, 64, 66, 131, 186
Receptive field (RF), 170, 179
Retinoic acids (RAs), 70
Ribbon synapse, 92, 168
RNA, ribonucleic acid, 8-10, 23
Root cells, 141
Round window, 94, 99, 101, 105-107, 110-112, 129, 172, 173

S

Saccule, 61, 69, 94, 108, 112, 130, 134
Salamander, 6, 23, 75, 78, 82-84, 87, 89-92, 101, 116, 129, 133, 171
Sall3, spalt like transcription factor 3, 63, 117
Sarcopterygian, 93-98, 101, 108, 171
Sauropsids, 100, 107, 173-175, 185
SC, superior colliculus, 153

Index **213**

Senses, 6, 33, 171
Sensory (cells, neurons) 34, 43, 47
Sensory epithelia, 2
Sensory receptors, 19
Sequencing, 5, 36
SGN, spiral ganglion neurons, 55, 93, 114, 116-120, 128, 130, 155, 167, 173
Shh, sonic hedgehog signaling molecule, 54, 58, 114, 119, 153
Singing, 176, 178, 186
Single cells, 2, 16, 20
Slc34a2, solute carrier family 34 member 2, 97
SOC, Superior olivary complex, 121, 138, 139, 144, 145, 149-154, 180, 186
Solitary tract, 65, 86
Sound pressure (processing), 93, 103, 112, 123, 173, 176
Sound, 4
Sox2, SRY-box transcription factor 2, 73, 87, 96, 97, 117, 124, 127, 153
Speaking, 178, 186
Speech, 176
Spherical bushy cells, 139, 143
Spiracle (duct, pouch) 98, 106, 115
Spoken language, 176
SPON, superior paraolivary nucleus, 141-149, 151, 156
Sponges, 17, 30-34
Srrm4, serine/arginine repetitive matrix 4, 68
SSA, stimulus specific adaptation, 157
Stapes, 101-104, 107, 185
Statocysts, 35, 37, 38
Stereocilia (stereovilli) 25, 40, 45, 53, 69, 89
Sturgeons, 82, 84, 89, 90
Superficial stellate cells, 142, 144, 145
Superior vestibulospinal track (SVST, SV), 62
Swim bladder, 95
Syllable, 177,
Synaptic ribbons, 89
Synaptic vesicles, 30
System, 4, 9, 12, 30

T

Tadpole, 99, 103
Taste, 52
Tau, neurofibrillary tangle protein, 183, 184
Tbx1-3, t-box transcription factor 1-3, 63, 68, 117, 125
Tbx4, t-box transcription factor 4, 96
Tectorial membrane, 93, 108, 110, 129, 173
Telencephalon, 91, 171
Teleosts, 74, 90, 95, 98, 104, 111, 122, 175
Tetrapods, 69, 70, 95, 97, 103, 107, 123
Thalamocortical, 164, 166
Therian mammals, 94, 111
Tlx3, t cell leukemia homeobox 3, 63, 68, 117
TMC, transmembrane channel like, 11-15, 70, 78, 79, 126
Tonotopic, 133, 140, 141, 144, 146, 155, 174, 186
Topographic, 162
Trachea, 95
Transformations, 122
Transmembrane, 12, 13
Trigeminal, 85, 171
TrkB, neurotrophic receptor kinase 2, Ntrk2, 63, 119
T-stellate cells, 139-140, 144, 148, 155
Tuberculoventral cells, 143-145
Tunicates, 56
Tympanic membrane, 98-101, 105-107, 112, 122, 129, 173, 185
Type I, 117, 118, 120, 125
Type II, 117, 118, 120, 125

U

Unipolar brush cells (UBCs), 137, 141-142, 144
Utricle, 61, 69, 108, 131

V

Ventral cochlear nuclei, 136, 144, 145
Vertebrates, 2, 6, 12, 34-36, 48, 70, 95, 115, 171
Vestibular hair cells, VHZ, 67-71, 86
Vestibular sensory organs, 28, 108
Vestibulo-ocular reflexes (VOR), 66
VGN, vestibular ganglion neurons, 55, 117, 119, 130

Visual, 171
VN, vestibular nuclei, 54
VNLL, ventral nucleus of the lateral lemniscus, 150-152, 156, 159
VNTB, ventral nucleus of the trapezoid body, 140, 145, 149
Vocal tract, 176
Vocalization, 176
Voltage-gated channels, 89, 127
VPL, ventral posterolateral nucleus, 160
VPM, ventral posterior medial nucleus, 160, 163

W

Weberian ossicles, 107, 129

Whales, 176
Wnt, 22, 54, 65, 76, 153, 154, 159

X

X chromosome, 6
Xenopus, 75, 78, 79, 87, 133

Y

Y chromosome, 6

Z

Zebrafish, 74, 78, 80, 102, 122